Preface

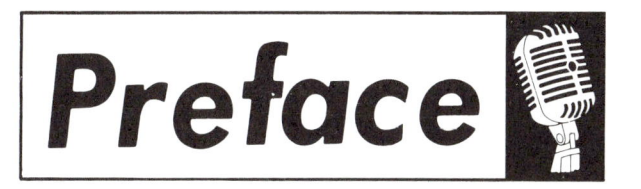

How many times, while watching a TV performer, or listening to a radio disc jockey, do you find yourself saying "How lousy! I can certainly do better than **that**!"

In many cases, you probably can do better—except that the performer is working as a paid artist and you are not. So, how is it possible for you, an unknown, to quickly smash down the invisible barriers that seem to encircle the broadcasting industry and break into the good-paying business of microphone and camera?

As a veteran of more than 30 years in broadcasting, I have learned and observed little tricks of the trade that will enable you to slip through a "side door" and get started in this glamorous and exciting profession. In these pages, you will learn some important career short-cuts that are time-tested and proven by the pros.

And best of all, you won't have to pay out a dime and you need not invest years of study to become a "name," earning money as a performer. Just read this book from cover to cover. Then select the approach that best suits your circumstances and apply its technique on an unsuspecting station manager in your area. The next voice you hear on a taped commercial may be your own.

My thanks to R.W. "Ozzie" Abolin for providing the photographs.

Sam Ewing

If you have half a mind to go into broadcasting, that's enough!

                    Graffiti

No. 620
$7.95

You're
On The Air

By Sam Ewing

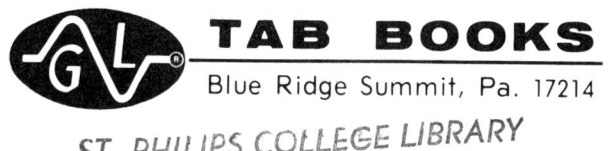
TAB BOOKS
Blue Ridge Summit, Pa. 17214

ST. PHILIPS COLLEGE LIBRARY

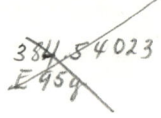

FIRST EDITION

FIRST PRINTING—NOVEMBER 1972

Copyright © 1972 by TAB BOOKS

Printed in the United States
of America

Reproduction or publication of the content in any manner, without express permission of the publisher, is prohibited. No liability is assumed with respect to the use of the information herein.

International Standard Book No. 0-8306-2620-4

Library of Congress Card Number: 72-87455

Contents

PART ONE: PREPARING YOURSELF

That Get-Ahead Ache   13
Your desire to succeed in a broadcasting career.

Station Breaks   13
Success stories of hopefuls who realized their ambitions.

Just Be Yourself   15
The key to success is to be yourself.

Style Is Your Stock In Trade   16
How to develop your "natural personality."

The Name Of The Game Is Selling   17
Why an announcer must be a salesman.

"Peddle Your Papers" With A Purpose   17
Learning broadcast salesmanship in your present occupation.

Profit By Experience   18
Involving yourself in commercials.

Home Study Made Easy   20
The importance of reading aloud and "growing a clock" in your head.

Watch What You Say!   22
Overcoming sloppy speech habits.

Tune Up Your "Engine"!   23
Voice exercises.

Develop Your Style   24
Putting yourself in the shoes of the individual listener.

"It Takes One To Know One"   25
The importance of studying the styles of big-time performers.

"Phone And Games"   26
How to rehearse your radio delivery by telephoning radio talk shows.

"Bleep" 4-Letter Words Out of Your Vocabulary   27
Why you should stop cussing and strengthen your talking machinery.

You Know, You Know, You Know   29
Avoiding pitfalls in everyday conversation.

Sweaty Palms and Paralyzed Larynx   30
How to overcome nervousness and knock out mike fright.

Talk Back To Yourself   33
Home practice with phonograph and tape recorder.

Give Yourself A Kick In The Seat Of The Can'ts!   35
How to avoid laziness in home study.

On Top Of The Action   36
Opportunities for fledgling broadcasters.

What About Broadcast Schools?   37
A look at schools of broadcasting, both residence and home study, along with a listing of universities, colleges and schools that offer instruction.

Terms Of The Trade   45
Terms of the trade and their definitions used behind the scenes.

PART TWO: GETTING YOUR START

Help Wanted—Everywhere!   49
How your first break lies just beyond the next bend in the road.

Think Small   51
Why small, local radio stations are the best training grounds.

Men For All Seasons 53
Reasons why versatility pays off.

Open Mike 56
Questions and answers on AM, FM, call letters, the FCC, station profits, station formats, and radio-telephone licenses.

A Typical Station 67
The typical small radio station. How it is staffed and operated.

Small-Station Radio 83
Operation of the grass-roots radio station with a minimum staff.

The TV Field 85
Opportunities in television.

A Typical TV Station 87
KNTV, Channel 11, San Jose, and how it typifies medium-sized operations.

Zooming In 89
A description of the physical makeup of KNTV.

Nice Work And You Can Get It! 95
Job descriptions of positions in a television station, and the qualifications required for getting them.

Talent Wired For Sight And Sound 134
Exploring in depth the potential for would-be performers in the cable (CATV) industry.

Ad Alley Avenues 137
Opportunities for performers at advertising agencies and program production houses.

Flesh Peddlers And Your Flesh 140
Talent agencies. Can they help you break into broadcasting?

PART THREE: HOW TO SELL YOURSELF

Happiness Is Being Johnny-On-The-Spot 141
How to arrange to be at the right place at the right time.

Segue To Success   142
Shortcuts to performance once you are inside the industry as a clerk or a sales person.

Dish Washer Today, Disc Jockey Tomorrow   143
You can change your place of employment and step into broadcasting.

The Personal Touch   145
The testimonial commercial as a direct route into radio-TV.

Unaccustomed As You Are   147
Using special knowledge and abilities to get your own show.

Air Breaks   152
How broadcast stations can use your talents.

Locker Room To Control Room   154
As an athlete of high school or college fame you can elbow your way into broadcasting.

Stage Flight   156
Amateur dramatics—a side entrance to on-air performances.

Public Exposure   158
Public appearances will help you into a radio-TV career.

The Model Employee   159
How a pretty girl can shape her TV future around a good figure.

And Now... The News   162
Breaking into broadcasting as a news stringer.

Audience Participation   164
The importance of appearing as a "professional guest" on talk shows.

It Pays To Win Friends And Influence Program Directors   168
Why contacts in the industry are so important, and how to cultivate friendships.

Ice Breakers 169
Secrets of dreaming up original program ideas that you can sell to broadcasting stations.

Play It Cool In Summer 175
Why the hot months are best for getting broadcasting jobs.

The Baby Sitter 175
How holding a First Class FCC Radiotelephone license gets you a radio position.

A Commercial For Yourself 177
Preparing a resume, the most important commercial you will ever deliver.

The Wrong Line 179
Hunting for work is a business. How you should approach it.

The Interview—Nods And No-No's 180
Do's and Don'ts in applying for broadcast jobs.

ABCs Of Auditioning 184
Three important rules to act by.

If At First You Don't Get In, Try, Try Again 184
Overcoming setbacks.

PART FOUR: CLOSING COMMERCIALS

Broadcasting As A Business 185
The better you can sell on the air, the more money you will earn for the station and for yourself.

Cold Cash And Hot Microphones 186
The size of paychecks you can expect in the radio-TV industry.

Powder Power 191
Opportunities for women in broadcasting.

The Sky's The Limit 193
How to start at the bottom and climb to the top of your chosen profession.

For The Love Of Mike And Camera   196
Why station executives love to spot a winner, and help newcomers along.

Something Going For You   199
Behind-the-scenes secrets of ad libbing.

Coffee Break Sales Seminar   204
Why you should know the techniques of selling, and a quick course in salesmanship.

Look Before Leaping   209
Pitfalls to avoid in your radio-TV career.

Sex And The Single Disc Jockey   212
Case histories illustrate the kinds of deep trouble in which a performer can find himself.

Pranks For The Memory   214
Practical jokers in the broadcasting industry.

Honesty Is The Best Policy Concerning Yourself   217
How to rate yourself as a potential performer.

INDEX   220

RECOMMENDED READING

Abernathy, Elton. FUNDAMENTALS OF SPEECH: William C. Brown, 1970. Practical methods of public speaking.

Allen, Steve. MARK IT AND STRIKE IT: Holt, Rinehart and Winston, 1960. An autobiography with recollections of Allen's radio career.

Barber, Red. THE BROADCASTERS: The Dial Press, 1970. Forty years of broadcasting are recalled by a famous sportscaster.

Buxton, Frank and Bill Owen. THE BIG BROADCAST: The Viking Press, 1972. A new, revised and expanded edition covering radio from 1920-1950. There are articles on announcers and disc jockeys.

Coddington, R. H. MODERN RADIO BROADCASTING: TAB Books, 1969. Management and operation in small to medium markets.

Dary, David. RADIO NEWS HANDBOOK—SECOND EDITION: TAB Books, 1970. A solid grounding in radio news basics, mechanics, and style, plus necessary details on the workings of a radio newsroom.

Dary, David. TELEVISION NEWS HANDBOOK: TAB Books, 1971. A complete guide to TV News department operation, including how to gather, write, produce and present TV news.

Elkin, Stanley. THE DICK GIBSON SHOW: Random House, 1971. This is a novel about an itinerant radio personality who is in broadcasting almost since its beginnings. (Available in paperback.)

Fisher, Hal. HOW TO BECOME A RADIO DISC JOCKEY: TAB Books, 1971. A comprehensive primer for the knowledge required of anyone seeking a career in the radio broadcasting.

Fisher, Hal. RADIO PROGRAM IDEABOOK: TAB Books, 1968. All the programming ideas you need to build and hold an audience! A thesaurus of ideas on radio showmanship.

Fisher, Hal. THE MAN BEHIND THE MIKE: TAB Books, 1967. This book offers practical and helpful guidance on every phase of announcing.

Higby, Mary Jane. TUNE IN TOMORROW: Cowles Education Corporation, 1968. An amusing account of an actress's career in radio. It recalls how many of today's leading stars got their start on soap operas.

Hoffer, Jay. ORGANIZATION & OPERATION OF BROADCAST STATIONS: TAB Books, 1971. An exhaustive examination of the responsibilities and capabilities required in each job classification.

Peck, William. RADIO PROMOTION HANDBOOK: TAB Books, 1968. Jam-packed with hundreds of ideas, and complete with scores of factual examples to spark hot, new ways of promoting a station.

Ris. PROMOTIONAL & ADVERTISING COPYWRITER'S HANDBOOK: TAB Books, 1971. Emphasizes the important aspects of preparing advertising or promotional copy for various media.

Routt, Edd. THE BUSINESS OF RADIO BROADCASTING: TAB Books, 1972. An up-to-date, comprehensive handbook on broadcast station operation and management—principally, how to operate a station as a profitable business and serve the public interest as well.

Siller, Robert. GUIDE TO PROFESSIONAL RADIO & TV NEWSCASTING: TAB Books, 1972. A practical self-study guide for those who want to get started or get ahead in broadcast journalism.

Swearer. COMMERCIAL FCC LICENSE STUDY GUIDE: TAB Books, 1971. Combines theory and applications with up-to-date questions and answers for 1st, 2nd, and 3rd Class Radiotelephone license exams plus broadcast endorsements.

Terrell, Neil. THE POWER TECHNIQUE OF RADIO-TV COPYWRITING: TAB Books, 1971. How to write broadcast copy that gets results, copy that will sell products and services.

Preparing Yourself

PART 1

THAT GET-AHEAD ACHE

For some time now you have been thinking that you would like to become a radio and-or television performer. I can understand that, for broadcasting is an exciting career, not overlooking the fact that it is a very profitable one. So what's stopping you from getting into the broadcast business? NOTHING—once you make up your mind to do it!

Age and appearance are no major handicaps, especially on radio where you are never seen. Lack of education is no real barrier. Many of the top stars in radio and television never finished high school. It takes all types of males and females to make up broadcasting—young ones, old ones, fat ones, thin ones, and in-between ones. If everyone had to be suave and slender like Bob Ewbanks, or beautiful like Barbara McNair, there would be no room for Orson Welles, Jackie Gleason, Phyllis Diller, Raymond Burr, Amy Vanderbilt, and Arthur Godfrey.

Accents do not work against you. Otherwise, the Galloping Gourmet and Eva Gabor would be out of business. Curt Gowdy, the colorful sportscaster, still has his pronounced Wyoming accent.

You can enter radio and TV from any walk of life. Dick Cavett was a copy boy. Carol Burnett, an usherette. Andy Griffith taught school. Julia Meade, Martha Hyer and Bess Myerson modeled clothing. Johnny Carson originally appeared as Carsoni The Great, a magician. Merv Griffin and Mike Douglas were dance band singers. Dennis Weaver delivered flowers. All of these people and many thousands of other who crashed into the magic circle once had an "inflammation of the wishbone" just as you have. They did something about it and made the grade.

Now it's your turn.

STATION BREAKS

Would you believe that a chubby, middle-aged housewife, with a voice like grinding truck gears and no broadcasting

13

experience whatsoever, could quickly find a niche for herself conducting a daily half-hour interview program, and make a decent living doing it? Off hand, you'd say "No chance!" Yet she accomplished it! Just one good example of slipping into broadcasting via a "side door."

First of all, "Mrs Hortense Alger," as I will call her, visited both radio stations in her hometown and suggested to the respective station managers that she be put on the air every noon-time interviewing diners in the town's leading restaurant. At both stations, Mrs. Alger was politely turned down by the managers who expected (and hoped) never to see her again.

But, in the best American tradition, she was a go-getter. Her next stop was the restaurant itself where she asked the owner if he would agree to a daily show, originating from his place of business, providing it cost him nothing.

"Sure," he said enthusiastically, "if it's for free."

Mrs. Alger then went from downtown merchant to merchant selling commercial announcements on the non-existent show. Within a few days, 15 merchants had agreed to each invest $25 per week for a 13-week period—a total of $4,875 to start with.

After her success in acquiring sponsors, the gal went to the radio stations and waved the signed contracts in front of the astonished faces of the station managers. Money talks! Mrs. Alger had both stations competing for her services, and, in the long run, she wound up making a deal with one of them to earn $600 a month for herself by hostessing what, for the past few years, in the opinion of veteran broadcasters, is the worst interview program in the history of radio. Except the listeners love it!

Among the more than 6000 radio and television outlets in the United States, not to mention several thousand cable television companies, many of whom originate local programs, there are many "side doors" to slip through. Across the country ordinary housewives like Mrs. Alger are starting on the air every day to build careers for themselves with cooking recipes and beauty hints, homecraft tips and other subjects of interest to the woman of the house.

TV has proved a natural for advocates of physical fitness. Jack LaLanne's body building program has made his name a household word.

Or take the manner in which radio-TV star Marvin Miller, "The Man of a Thousand Voices," got his start years ago. It was during the Depression. Marvin wrote a radio dramatic show for seven characters, and he auditioned all seven parts himself. A St. Louis radio station hired him immediately on the sensible theory that one actor was cheaper than seven. By the time the program really got rolling, the cast was up to 42 characters, and Marvin Miller was still the only actor on the payroll. Just as that old saying goes, "There's more than one way to skin a cat," and there is more than one way to break into broadcasting.

JUST BE YOURSELF

In the early days of radio a big deep voice was essential for an announcer. Every young fellow who aspired to talk on the air professionally wanted to sound like Harry Von Zell of the "The Fred Allen Show," Westbrook Van Voorhis, the voice of "The March Of Time," or Singin' Sam the Barbasol Man. What's more, the men who did the hiring wanted booming voices.

Thus, in 1936, a thin-voiced 16-year-old kid, with ambitions to become a "voice" on WQBC in his hometown of Vicksburg, Mississippi, slept with his bare feet exposed next to an open window. He caught a terrible cold that brought his voice down a couple of octaves. The next day the stripling read for an audition at the radio station and was hired on the spot as a part-time announcer.

Of course, the kid almost died of walking pneumonia, but he did break into broadcasting by applying a dramatic shortcut. Once his cold disappeared, however, his voice returned to its normal squeekiness, and the station manager wondered why he had hired the kid, but didn't have the heart to fire him.

As the young fellow grew older, his voice naturally deepened, and, in time, he sounded professional enough to get important jobs on the Pacific Coast. Eventually, he gravitated into television as a writer, producer and "voice over" for national shows and commercials. I know that kid well. It was myself.

Luckily, these days no one needs a "thunder throat" to make the grade in broadcasting. **Naturalness** is the key ingredient for selling goods and services to listeners. The secret is to be **yourself.**

Certainly, no one thinks of such prominent personalities as Ed Sullivan and Lawrence Welk as stereotyped hosts. Yet both of these gentlemen found fame and fortune just by being themselves.

And there's Dizzy Dean. The former Cardinals pitching star made a huge success as a sportscaster of baseball games with a naturalness that made English teachers cringe. "Slud into second" was one of Dizzy's typical language innovations.

On one particular afternoon, the old pro startled his transcontinental audience by coming up with, "The trouble with them boys is they ain't got enough **spart**." Asked later for an explanation of this, Dizzy obliged: "Spart is pretty much the same as fight or pep or gumption. Like the **Spart Of St. Louis**, that plane Lindberg flowed to Europe in."

Another baseball great, Willie Mays, once showed up for a role as himself in a television show. The director asked how he planned to play himself. "I don't know," shrugged Willie. "Just turn those cameras on, and if it ain't me, let me know."

The point is, you must work hard at just being yourself. For, once exposed to the general public, your own collection of funny little habits can pay off beautifully on transistor and tube.

STYLE IS YOUR STOCK IN TRADE

Just as a plumber can't work without his plungers, wrenches or snake, a broadcaster must have handy a kit of tools to succeed in his trade. In broadcasting, we call it style.

What is Don Rickles without his insults, or Julia Child without her recipes? Jonathan Winters is lost without his cast of kooky characterizations, or take the obvious equipment of Eva Gabor. Each performing artist has to determine at the outset the style that most naturally fits himself or herself in order to become an air personality.

Before going any further, let us decide on your's. Check off below the type of personality that you believe bests suits you:

BREEZY & BOUNCY? ( )

CALM & AUTHORITATIVE,
UNRUFFLED BY ANYTHING? ( )

FAST-TALKING WITH A SPEAKING
STYLE MUCH LIKE A CARNIVAL BARKER? ( )

A SINCERE & WARM PERSONALITY
SPEAKING AS ONE NEIGHBOR TO ANOTHER
OVER THE BACK FENCE? ( )

WITTY, LIFE OF THE PARTY? ( )

SOUR, DRY WIT? ( )

No matter what you are trying to peddle on radio or television, the first thing you must do is sell yourself. But in order to **sell** yourself, you must make sure you know just what it is you are, and develop that natural peraonality just as a weight lifter builds up his muscles sinew by sinew.

THE NAME OF THE GAME IS SELLING

One evening a television network sales executive carried out a little experiment. While watching a championship fight on his network, he responded to all the commercials, as he hoped the millions of viewers would likewise do. Each time the announcer urged him to to to the refrigerator for a can of beer, he did so. Drinking furiously, he managed to stay even with the announcer, guzzling a fresh, cold brew each time he was invited to. As a result, the losing boxer was knocked out in the 11th round. The executive was flat on his back by the end of the 10th.

Moral: The broadcast media has to sell products to stay in business, and a good on-the-air pitchman can always find a slot in the industry.

"PEDDLE YOUR PAPERS" WITH A PURPOSE

Nobody is a "born" radio or TV pitchman. The art of squeezing the last drop of sales from aspirin, eye liner and underarm deodorants in the mass media must—and can—be learned.

Now let us take a look at your present occupation. Out in the workaday world, chances are you are in training as a media pitchman right now and don't realize it because your line of work seems so far removed from broadcasting. But let's analyze just what you are doing. If, for example, you are selling sacks of bird seed, lingerie or bakery buns, electrical appliances, used cars or cosmetics, you are, in fact, preparing yourself in the best possible manner toward becoming an on-the-air sales person. There is no better experience!

In the broadcasting business, as in all walks of life, it is sales that make the world go round. The objective is to **move merchandise** for advertisers. Otherwise, without the income of almighty dollars, no commercial station can survive.

Thus, when you face a prospective customer, who is looking over your display of vacuum cleaners, stretch socks or barrel of apples, you have to think of that individual as **an audience of one. Your sales talk is your script.** When you sell your product to him, you score just as successfully and no differently than John Cameron Swayze hawking waterproof watches on coast-to-coast TV. His audience is larger, that's all.

The handsome TV announcer, Lyle Waggoner, best known for his appearance on THE CAROL BURNETT SHOW, credits door-to-door peddling of encyclopedias, health club memberships and a wide variety of other products, back in his hometown of St. Louis, as perfect on-the-job training for his later profession of pitching merchandise on the air.

Ed McMahon, Johnny Carson's foil on THE TONIGHT SHOW, and sometimes MC of daytime network game shows, was a bingo announcer in a carnival and a boardwalk pitchman in Atlantic City before getting his on-the-air start in Philadelphia. His early spiels to the crowds were great training for a reported $250,000 annual income.

Hawaii Five-O's Jack Lord, during his days as a struggling performer, picked up eating money selling cars in New York.

One executive of the Fuller Brush Company has stated: "All salesmen have to do a little acting. In this area, we get a lot of applications from actors. Some of them are very good salesmen; they know how to talk to people."

As a bank teller, paint salesman or restaurant waiter, your best means of getting experience is to pretend that each customer is an audience of millions. Put your heart into your pitch. In the long run, if you actually do perform before an unseen audience that runs into seven figures, it will be duck soup for you. And, incidentally, in the meantime, your sales will definitely pick up in your present job while you are "rehearsing."

PROFIT BY EXPERIENCE

Experience in selling is helpful in more ways than one for an announcer. On local radio stations, often is heard a half-hearted word on behalf of an advertiser. It is obvious that the announcer is merely mouthing phrases that have been typewritten on a piece of paper and handed to him.

Nobody is getting a fair shake! The merchant who has bought a minute's time in which to motivate listeners to rush in and buy his merchandise is not getting his money's worth. The broadcasting station is losing, because the public will not be inspired to buy from the advertiser, and the latter will cancel his current flight of spot announcements, probably never buy a schedule again, and he will "bad mouth" radio advertising in general.

The anouncer is the third and largest loser because, by letting both the sponsor and the station down, he has also let himself down badly. Since he tosses off commercials listlessly, he brands himself as a small timer who will always remain in the "small time." On the air an announcer must involve himself in each commercial message as if he were the advertiser himself facing a thousand prospects in a large meeting hall. Think of each commercial as money in your pocket, or, better still, as your best girl and you are kissing her goodnight. Put warmth and feeling into each message.

Jimmy Wallington, one of the highest paid announcers in the business, once told me that he took special care to always emphasize the name of the product as if it were the greatest thing in the world. If you listen carefully to the big-time announcers in radio and TV, you will discover that they all use this simple technique.

In the beginning it took me a few years to realize how important it is to **sell** each announcement with feeling. My teachers were hard-bitten businessmen out to make a buck. Originally, as a disc jockey (except we weren't called deejays in those days), I read whatever was placed in front of me with the best diction and most enthusiasm I could muster, while doing my best to speak from deep within my chest. I was striving for **mechanical perfection.** Yet I was missing the real boat. I was not **involved** in the products themselves.

The time came in my early career when a need for extra money to support a wife, four young sons, a dog and a Rhesus monkey inspired me to try selling radio advertising to supplement my income. Following my 6 AM to 2 PM broadcast shift each weekday, I hit the street and made calls on downtown businesses. My deal with the station manager was 20 percent commission on any advertising I sold.

Right off the bat I learned the serious side of commercial radio. I found that no merchant is satisfied with anything less than immediate results. And if he "tries" radio and it fails to produce results for him, you have had it for good.

So, please picture me peddling a schedule of spot announcements to the man at Blain's General Store to sell out a warehouse full of what we politely called "steer guano."

"I'll gamble for 37.50," he said. "But I doubt you can do us any good." From his pessimistic attitude I knew that any future sales to Blain's and commissions for myself would be non-existent unless he sold those blasted sacks of fertilizer.

Now, I had a **personal interest** in doing a selling job. The next morning when I read on the air, "Giant Blain's General Store has genuine steer guano for just 99 cents a bag—the steer's loss is your gain with Giant Blain"—believe me, I put my heart into it. I thought of my $7.50 commission on the small sale, and the pair of shoes it would buy for one of the children. I thought of the man at the store grumbling to his bookkeeper that he had flushed $37.50 down the drain. I thought, "Dammit, I'm going to sell this stuff."

As I read the words with feeling, in my mind's eye I drew up a vivid picture of all those smelly sacks resembling bags of gold, and I visualized a virtual army of farmers breaking in Blain's door to buy them. And you know what? Thanks to my **personal involvement** in the commercial, Giant Blain's sold every blessed bag!

HOME STUDY MADE EASY

Read the following 30-second test commercial aloud:

LADIES...HAVE YOU TRIED SCHNOOK'S SUPER SOAP, THE NEW WASHDAY MIRACLE? SCHNOOK'S IS MADE EXCLUSIVELY FOR OLD-FASHIONED FOLKS WHO BELIEVE THAT CLEANLINESS IS NEXT TO GODLINESS. WITH SCHNOOK'S, IT'S NEXT TO IMPOSSIBLE. SCHNOOK'S...THE SPECIAL AMMONIATED, HOMOGENIZED, BIO-DEGRADABLE, DETERGENT SOAP POWDER COMES IN THREE CONVENIENT SIZES—SMALL, MEDIUM AND LARGE, ECONOMY FREIGHT-CAR SIZE—JUST PERFECT FOR WASHING FREIGHT CARS! REMEMBER SCHNOOK'S SLOGAN: IF YOU DON'T USE OUR SOAP, BE SURE TO USE OUR DEODORANT!

When you are able to read the comic Schnook's Soap commercial aloud in exactly half a minute, without fluffing and with good inflection, you have jumped a giant hurdle.

An announcer should be able to sight read almost any piece of copy that's handed to him with confidence and sincerity and within the designated time limits of 10, 20, 30, and 60 seconds. A professional grows what he calls "a clock in his head." By that we mean he can pick up an unfamiliar piece of copy and read it "cold" in the specified time period.

Practice is the secret. And you are going to have to read aloud millions of words to achieve your ultimate goal. But, when your chance comes to perform on the air, you will be prepared. Starting today, and for at least an hour every day from now on, you will want to train yourself for broadcasting by **reading out loud.**

The daily newspaper is one of the best sources of practice material. It is always fresh and contains greatly diversified styles of writing. It covers everything from the comics to dramatic stories on world events and advertisements on everything from snappy girdles to green houses. You can practice anywhere—during coffee breaks and lunch hours, on a bus, and so on. Anywhere and everywhere you have some free minutes.

Even if strangers look at you like you are some kind of nut, why not sacrifice their quizzical stares in the interest of achieving your ultimate goal? Think of it this way: Some day these very same people are going to tune in specifically to hear you. Why not give them an in-person prevue of things to come? Besides, in the future you will have to work in front of audiences one way or the other, and you must overcome any self-consciousness. Get used to the stares of strangers now!

Whenever possible, shut yourself in a room by yourself and read the label on a can of soup, instructions for assembling a musical bird bath, chapters out of a sex manual. I am even so bold as to suggest that you read YOU'RE ON THE AIR out loud from cover to cover. After all, it is written in a commercial, narrative style rather than in the jargon of the technical textbook.

It doesn't matter what you read, so long as you practice **inflection** and **timing**. And do this every day, just like singers practice singing. Glen Campbell, Pearl Bailey, and Andy Williams didn't just pop out of a toaster. They worked hard to hone their vocal skills, and so must you!

You will no doubt flub and get your tongue twisted. Everybody does it, even the pros. Take the case of the network announcer who was describing the arrival of the governor of the Virgin Islands to Washington. The announcer who made the unfortunate error garbled his words and said into the microphone: "Today the White House is to have a special guest—the Virgin of Governor's Island." Despite your own fluffs, you'll be surprised how quickly your delivery will improve.

It's important to use a tape recorder and speak your lines into it. The playback is a severe critic and will tell you instantly the things you are doing wrong.

Once you have mastered the technique of professional reading, memorize the copy of a newspaper or magazine ad and start working in front of a mirror. Again, your reflection will show you your mistakes.

After you have worked at your style and delivery for several weeks and feel that you have achieved marked results, try reading to members of your family or close friends. If, after listening to you, they remark that you have a "nice voice" and it lulls them to sleep, or if, on the other hand, they grimace and groan, then you had better practice a great deal

21

more. But, if you can inject a spontaneous, conversational quality into your "commericals" and they ask you to repeat them, you are making excellent progress.

Stated briefly, you will find it pays rich dividends to talk to yourself—and others!

WATCH WHAT YOU SAY!

Steve Allen once presented a humorous skit in which he demonstrated with flip cards how Americans are creating a new language. "Gunagetcha," according to Steve, is a common expression meaning "going to get you," whereas "Zeeyalader" means "see you later." His point was, that, for the most part, our speech habits are sloppy and we run our words together unnecessarily.

As a professional announcer you have to break yourself of bad habits like these. If you tune in the networks, and listen with a critical ear, you will notice that the pros articulate carefully. The thing to do is emulate their good diction.

The most common slurring heard across the United States is as follows:

SHUDDA for SHOULD HAVE

GUNNA for GOING TO

DONCHA for DON'T YOU

The letter "W" should be articulated as DOUBLE-U, not DUBYA.

It's GET, not GIT.

JUST instead of JEST.

BECAUSE and not KUZ.

Don't use YER for YOUR, TAH for TO, YOUSE for YOU, DAT for THAT, DOZE for THOSE, or 'EM for THEM.

And never call it NOOZ. NEWS is pronounced as if it were spelled NYOOS.

If a word has an ING at the end of it like DOING, say DO-ING and not DO-IN. Say GO-ING and not GO-IN.

Never put Rs on words where there are none. It's not CUBAR, it's CUBA.

Don't swallow words. Pronounce all the Ts and Ss.

The region in which you live may be responsible for many of the verbal ruts into which you have slipped. New Englanders are overly generous with the broad A. In the West, the natives substitute A for O, and consequently, BOND becomes BAND. A Southerner drops his R in pronouncing SIR as SUH. Regardless of the reasons for your present imperfections of speech, it is necessary you correct them.

But, getting back to network announcers, please notice that it is almost impossible, from their manner of talking, to determine from which part of the country they come. That's because they have practiced diligently to establish a standard pattern of speech.

I think that it is no exaggeration to say that 99 percent of all the people you meet mispronounce hundreds of words. Chances are you are in the same bag, so start working with a pronunciation dictionary right away. A small, vest-pocket type is the easiest to use, since you can carry it around with you for instant reference.

As a broadcaster, you must pronounce words correctly. Otherwise, listeners tuned in out there in cars, in beds and on beaches will think of you as a prize dummy. Remember that the microphone is a highly delicate instrument, almost human in sensitivity, and it magnifies everything about your speech and personality. Other than your loved ones, nothing gets as close to you as a mike does.

Conclusions:

(a) **Listen to network announcers closely. Pay attention to their pronunciations and imitate them.**
(b) **Check your dictionary constantly.**

TUNE UP YOUR "ENGINE"!

Have you wondered why singers always warm up before a performance? Believe me, they don't do it for kicks. They exercise their vocal cords and muscles because it loosens them up.

The singing machine has to react faster than the conscious mind. In other words, if the vocal apparatus doesn't react properly, a singer will sound flat and off-key and his voice will crack. It's the same with announcers. Their voices have to be exercised, too.

At this moment, your voice is probably in need of a lot of work, so, like a singer, you will have to exercise your vocal engine regularly to strengthen it and increase its vocal range. To start with, you never want to shout or strain your vocal muscles. You want to treat your voice kindly, because in broadcasting it will be your most valuable tool.

When you wake up every morning, tune up your speaking machinery. Begin by taking a deep breath and say UHHHHHHHHHHHHH, until you run out of breath. Try to get the UHHHH sound as deep in your chest as possible, very, very low. Next, do the same with AHHHHHHHHHHHHH.

Devote two minutes to the UHHHHH exercise and two to the AHHHH. Within one week, you will notice an amazing improvement in your voice.

At the corner of Hollywood and Vine a tourist in town once asked Lee Giroux, the network personality, "What's the best way to get to ABC?"

"Practice—and lots of it!" was Lee's answer. And practice it is.

It is most important that each and every day, without exception, you practice your reading, putting particular emphasis on enunciation and pronunciation. **Then start over and do it eleven times more!**

DEVELOP YOUR STYLE

The engineer stabbed a finger at me. In turn, I signaled the short, bald orchestra leader who waved his fiddle bow and the band struck up the theme. Cupping a hand over my left ear so I could hear myself over the blasting brass and strings, I plunged into my opening bit:

"From the beautiful Blue Room on the Sunset Highway, just east of the historic city of Vicksburg, we send you the unique rolling rhythms of Goetz Granville, The Little Maestro, his singing violin and his orchestra...!"

That's the way announcers sounded on radio back in the late 30s, and, in those days, it was always my feeling that each announcer was trying to out-shout, out-enunciate and out-orate the other. In general, our styles were exaggerated, bullying and far from conversational. Radio had not become the **intimate medium** that it is today.

Unlike the commentators of 30-odd years ago, today's radio and television performer must not adopt the frenzied tones of the orator seeking to move millions, or the patronizingly superior announcer on the dance band remote booming over trumpets and trombones. Even though you may be addressing an audience of thousands, even millions, you must keep your efforts scaled to a single person who may be listening or watching all by himself in a lonely room.

Ask yourself: Who is this individual? What are his interests? His tastes? It is the accuracy of your judgment, based upon the type of material you are presenting, which will, for the most part, determine your success.

By way of illustration, if you are presenting the evening news, you are giving out information to people who are interested in knowing just what is happening in the world. This is the time to be concise and present your news stories in a clear and logical manner. No one wants you to be a great dramatic artist or another Williams Jennings Bryan.

On the other hand, if you are pitching used cars for Honest Harold, the dealer with a heart, you will get much farther if you show, at least by inference, a real sympathy with the hapless driver who chugs along everyday in his 1965 clunker. You simply outline the superiority of your product and try to excite the buyer to action.

These are only two examples to illustrate the point of **intimacy** in broadcasting. The fellow interested in the used car requires information, but he also needs the sympathetic attention of someone who has an interest in his problem. If, as a newscaster, you tried to apply the same slick approach, you would not be a factual reporter but a propagandist.

You must strive to understand the significance of the material you are presenting. Don't get absorbed with the mechanics of your vocal process, listening to the sound of your own voice, but think of **what you are saying, and to whom you are speaking**. Everything about your delivery must give the listener or viewer the feeling that you are confident of your product's ability to live up to claims made about it.

A good rule to keep in mind is: **Always be believable. Put yourself in the shoes of the individual listener and-or viewer at home.**

"IT TAKES ONE TO KNOW ONE"

One of the best ways to learn to become a professional performer is through studying the more successful examples around you. They are everywhere, and you will be surprised at how quickly the principles of performing become evident if you know what to watch and listen for.

Take a good look at Walter Cronkite on the CBS EVENING NEWS and study his style. Tune in Bill Cullen on the radio, and ask yourself what makes him so popular. Listen critically to announcers who "voice over" copy on national television commercials. Put your favorite disc jockey under the mental microscope and examine his good points.

In each case, you will find that these people establish a **rapport**, that is, a close relationship, with their audiences. The public likes them, and comes to think of them as close friends visiting in their homes.

For years Arthur Godfrey has shone brightly as radio's best salesman. On the air he sells merchandise by trainloads for his sponsors. This is due to Godfrey's neighborly approach and one that he developed by accident.

Like most of us in the early days of radio, Arthur Godfrey was an unspectacular announcer speaking into the microphone, in a stilted, preaching manner. Then, for a period of time, he was confined to a hospital bed where he spent most of his day listening to the radio. He started analyzing announcers, trying to figure out why one broadcaster was superior to another.

At last he came to the conclusion that the **conversational** quality was lacking in the media and that an audience should be talked to like one friend speaking to another on the telephone and not like a corporation president delivering an annual report to the stockholders. As soon as he was back at the microphone, Godfrey experimented with his new formula. It was an instant success, and Godfrey parlayed it into a fortune.

In short, the audience has to like the announcer to believe in what he says. Add to this the basic ingredients of relaxed delivery and good diction, and you are "in."

"PHONE AND GAMES"

In many cities, the radio "talk" show has caught on. And some stations, such as KGO Radio in San Francisco, schedule nothing but conversation programs 24 hours a day, seven days a week. In most cases, unidentified listeners phone in to discuss a variety of subjects—taxes, pollution, racial problems, transportation, unidentified flying objects, witches and warlocks, putting bells on pussycats, and so on endlessly. You name it; it's talked about!

Without callers, the program host, usually called a "communicastor" is up that proverbial creek without a paddle. He cannot fill his 3- to 4-hour air stint without continuing response from the general public.

Here is your golden opportunity to rehearse your embryonic radio style, while enjoying some actual on-the-air experience and helping the communicastor keep his ratings up. Think of a good subject about which you are knowledgeable and call the man at the station. If your approach is intelligent, you can become a regular caller and an unidentified "name" in a small way. Remember, one thing leads to another.

One of the earlier talk shows was conducted on KGO in the early 60s by Les Crane who broadcast from a now defunct San

Francisco nightclub known as the **Hungry i.** One of Crane's regular callers in those days was a sportscaster named Ira Blue whose distinctive voice and speaking style became immediately recognizable to listeners other than the sports minded. Thus, when Les Crane resigned his KGO post to host a national TV program, it was Ira Blue who was chosen to replace him at the microphone. And over the past years, Ira has grown in stature as a communicastor to become the dean of talking hosts on the West Coast.

"BLEEP" 4-LETTER WORDS
OUT OF YOUR VOCABULARY

Profanity has no place in broadcasting, if anywhere. Although 4-letter words are greeted with interest and sometimes applause in the movies and on the stage, if a television or radio announcer uses one he will probably be fired forthwith. Such "air pollution" is not tolerated in the industry.

An historic slip that slowed up a brilliant career was made by Don Carney on his "Uncle Don" radio program that once brought entertainment to the kiddies from WOR in New York. Don thought his microphone was cut off when he ad libbed a remark at the end of his show: "I guess that will hold the little bastards!" On that terrible day, Uncle Don learned that an announcer must make sure he is off the air before voicing any private comments. But in his case, it was a lesson learned too late.

Words and phrases often used in private conversation inadvertently find themselves going out on the air waves. NBC-TV personality Joe Garagiola, a former baseball player in the major leagues, in late 1971 came on sounding like he had let the old team down. "Jesus Christ, I'm sorry, goddamit" he cried after repeatedly blowing lines while taping of all things, a Christmas commercial. A studio technician by mistake included the profane segment in the tape of the TODAY show and it was unwittingly broadcast on more than 100 stations from coast to coast.

Red-faced, I confess that a few years ago while doing a radio commercial for Schmidt's Sheet Metal, in which the advertiser's name cropped up a dozen times, I eventually blew the word "sheet." Nineteen telephone complaints from alert and angered listeners, including Mr. Schmidt himself, reminded me of my blooper.

Unintended indiscretions are bound to happen on **live** radio and television despite safeguards. Take the popular

27

telephone talk shows. Radio engineers build in a special tape system by which all conversations on the air are delayed approximately six seconds. By holding down a button, the program producer bleeps out any profanity used by the caller. Thanks to this electronic method, you do not hear just where it is that an irate listener sometimes tells the host to shove his program. Nevertheless, profanity sometimes sneaks through because the producer's finger is not quick enough or there is some sort of mechanical failure.

I am reminded of an experience suffered by a staff member of a small 250-watt station in Indiana. On Sunday mornings, he would play an electrical transcription of a sermon by a hell-and-brimstone evangelist on a turntable while he stepped into a restaurant next door for coffee and a sweet roll.

The announcer had everything perfectly coordinated. He was always back in the control room within 12 minutes, allowing himself plenty of time to make his closing announcement, identify the station and get the next program underway. But one Sunday morning he returned to the unattended station to find the telephone flashing angrily, all three lines were ringing. Early in the program, the needle had stuck in the groove of the ET, and for ten minutes listeners heard nothing but the preacher's words, "Go to hell, go to hell, go to hell."

It is my belief that a performer can avoid embarrassing and dangerous slips of the tongue 99 percent of the time if he himself bleeps profanity out of his everyday speech. In other words, DON'T CUSS, **even in private!** Four-letter words are used by poor talkers to fill in blanks in their conversation when their brains are missing fire.

To resort to profanity will weaken your vocabulary, and you will come to rely upon expletives instead of seeking the word that will express your thought correctly. Just think how important it is to be able to grasp the right word at the right instant when you are conducting a man-on-the-street interview, broadcasting a sports event or hosting your own telephone talk show or quiz contest. You won't need to rely on UHs and AHs as a crutch while groping for the proper action adjectives to replace the customary 4-letter words that automatically pop into your head.

The microphone is a sensitive instrument and can be dangerous to your financial health. Thus, Rule One for better ad libbing: Stop cussing and start to strengthen your talking machine. Profanity is a way of escape for the man who runs out of ideas.

YOU KNOW, YOU KNOW, YOU KNOW...

How often do you hear people saying, "You know?" This expression, so meaningless in most usage, just has to be the most over-worked phrase in the English language! Used in everyday conversation, "you know" sometimes pops up as a question: "KSFO broadcasts the Giants games, you know?" Often as a statement: "You know, I think Sandy Duncan is a very fine actress." As a crutch for sentence punctuation: "You know, I want to break into broadcasting. It's a good field, you know. And, you know, I'm sure I can make the grade. You know?"

Stop for a moment and play back in your head some of your recent conversation. If you are an average citizen, chances are you continually pepper your speech with worthless "you knows" without realizing it.

Now the time has come for you to take action and dress up your vocabulary. A professional announcer drills himself to avoid all the "you knows" and other moth-eaten slang phrases. If he did not, he would sound like a cracked phonograph record with a needle stuck in a groove.

Part of the basic equipment needed by every announcer is a proper choice of words. And **word power** demands:

(a) Deleting from your vocabulary common slang that has become cobwebby ("uptight," "out of sight," "would you believe," things like that). Lots of people, young fellows particularly, riddle their speech with "man," definitely a banal expression that you must scuttle.

(b) The learning of dozens, hopefully hundreds, of new words and their precise meanings in order to use them fluently in general conversation, as well as on the air.

Three simple steps, methodically followed, will help you achieve a fluid and effective vocabulary in just a few weeks.

Step One: Look and listen for new words. You must discipline yourself to be on the lookout, in your reading and listening, for words that other people are using and you are not.

Step Two: Add these words to your vocabulary. Newspapers, magazines and books are filled with words that are unfamiliar to you. While you are reading aloud, practicing your diction, delivery and style, and come to a word with which you are unacquainted, do not pass it over impatiently. Instead, pause for a moment and say it over to yourself; get accustomed to its appearance and sound. Look up its exact meaning in your dictionary. As a result, you will begin to notice how this same word appears unexpectedly again and again in all your reading. After you have seen it a few times, your mind is set for it. You will know exactly what it means and you can use it yourself.

....**Step Three: Set a goal!** Plan to add five or ten new words to your vocabulary each day. This may sound overly ambitious, but, as you will discover overnight, it is an easy program to follow and your vocabulary will build like a snowball.

As you systematically increase your word power, you will also sharpen and enrich your thinking. New words will help to build your self confidence. They will improve your ability to express your thoughts more effectively, and help you to come out on top in the broadcasting business. You will speak easily and without embarrassment—without "you know" as a creaky crutch—and you will be a better person for it. Speech is the expression of one's personality, and you want yours to sparkle (you know)!

SWEATY PALMS AND PARALYZED LARYNX

Comedian Dom DeLuise has a very funny routine in which he portrays the representative of a savings and loan association who is about to appear for his company on live TV.

In rehearsal Dom makes an exceptionally smooth sales pitch for the sponsor. His spiel and physical action are perfectly timed, and his entire delivery is impressive. However, the instant the on-the-air sign flashes on and the floor manager cues him, Dom is seized by "camera fright," and his presentation falls apart. He stutters and stammers, attempts to light a fountain pen instead of a cigar, tangles himself in a prop telephone cord and blows the commercial completely.

As is true of most comedy acts, Dom's hilarious bit is based on actual happenings, and I am sure that his routine brings cold chills to lots of professionals who once experienced paralyzing terror during their early appearances at the microphone or in front of the camera.

After more than 25 years I still recall with reincarnated horror how I was suddenly rushed into a studio to interview the Honorable Earl Warren, then Governor of California, and how I froze at the microphone. My tongue became like a piece of petrified wood, and I could think of nothing at all to say. Luckily, another staff announcer, Jack Murphy, bless his heart, came to my rescue like a cavalry charging over the hill. Otherwise, I'm afraid, there would have been long moments of dead air on KSRO, Santa Rosa.

At any audition or first performance, you can expect to be keyed up. When your moment comes to speak, you may discover that your tongue has gone dry and that you have an overpowering urge to rush to the restroom. Nervous coughing and repeated blowing of the nose are likewise natural reactions.

When nervousness strikes, the mental machinery simply does not function as well as it does normally because anxiety undermines your efficiency. Such a nervous condition is probably the most powerful nullifier of talent that exists in a newcomer.

No one is immune. Often seasoned veterans of broadcasting are gripped with camera or mike fright, especially when thrust into an unfamiliar situation, as I was in the Earl Warren interview, or for a strange variety of causes.

Commentator Lowell Thomas has been quoted as saying that even after decades at the mike he still feels a certain tension just before he goes on the air.

"Funny Face" Sandy Duncan, a graduate of numerous TV commercials, and whose work before the cameras is always outstanding, complained to one national magazine reviewer that her hands were always cold before a scene because she couldn't control her nervousness.

A top-flight national personality, whom I will not further embarrass by naming here, was caught in a "Dom DeLuise" situation of a live nationwide sports telecast, when, during his introduction, it was mentioned that he would be speaking to 30 million viewers. The thought of this multitudious number of persons watching obviously shocked him. He became speechless, his hands trembled, and he spilled the beer he was supposed to pour and speak the praises of.

In Iowa a usually glib announcer hosting a man-on-the-street remote program unexpectedly faced a famous movie actress who was ushered to the microphone by an assistant. The announcer paled and words failed him. Finally, he managed to croak: "That's it from here. We now return you to our main studio." Controls were transferred seven minutes earlier than scheduled.

I remember a newspaper editor who used to report the noon news on a radio remote from a little studio in his editorial offices. This man's voice and delivery were excellent, and the news was professionally presented by the editor in a calm, relaxed, well organized manner that earned for him a large tune-in every Monday through Friday.

Oddly enough, the editor insisted, as part of his deal with the broadcasting station, that he remain anonymous and never be introduced by name, although every listener in the county knew his identity. One day, just for the hell of it, a studio announcer, upon switching controls to the newspaper offices, identified the editor. At first, there was a long pause. This was followed by several gasps and a gulp. Then came the voice of the editor in shaky, frenzied tones. A third of the way through the newscast, he broke down completely. He stopped reading and cut off his microphone. To my knowledge, he never spoke on the radio again.

Perhaps a psychologist can explain this hang-up. I will not attempt it, but merely point out to you, as a performer-in-training, that you have to find ways and means to overcome such a problem. An announcer must have confidence in himself, just short of outright egotism. He has to be poised and authoritative; audiences are quick to detect nervousness and uncertainty.

Each of us who appears professionally develops a personal gimmick by which we face each new piece of continuity, interview or unusual situation with "cool." The most common method that working announcers employ for relaxed delivery is by thinking of a particular individual in the unseen audience (a wife, sweetheart or good friend) and talk directly to that person. A prominent network commentator once admitted that, in his mind's eye, he always envisioned a network vice president sitting on the john. This thought, he chuckled, invariably soothed his nerves.

Winston Churchill's advice to speakers can certainly be adapted to an unseen audience as well as any live one that the British Prime Minister faced. The great Churchill advised: "Don't be nervous. Do just as I do. When I get up to speak, I always make a point of taking a good look around the audience. Then I say to myself, 'What a lot of silly fools.' And then I always feel better."

Relaxation is the key to a smooth performance, and in achieving this effect I recommend a type of mild, self-hypnosis that brings about complete body relaxation. Before going on the air, you should close your eyes and allow your neck and shoulder muscles to be as loose as possible. Say to yourself, "I am becoming more and more relaxed...more and more relaxed." You repeat these suggestions over and over again while you rotate your head five times in a clockwise circle, letting the muscles loosen still more. Then reverse the movement, making it counterclockwise for five more times. This entire exercise takes less than a minute and works wonders in calming you.

Deep breathing just before going on the air is another technique helpful in overcoming nervous tension as Yoga practitioners have discovered over a period of 2000 years.

Simply inhale deeply, hold your breath for a few seconds and then exhale. Try this exercise a few times. You will instantly notice the calming effect.

Sometimes even very large yawns will achieve relaxation. Yawning, however, is the least recommended method, inasmuch as people around you may think you are disinterested in what you are doing.

Whichever method you choose, you must learn to relax. Nervousness or tension during those first precious moments in a studio are bound to affect your performance to its detriment.

Yet there is a consolation to bear in mind: **The auditioner, contrary to the general opinion of hopefuls, is rooting for you 100 percent and wants you to be good. He is on your side and will give you every break and consideration possible. This thought in itself should relax you.**

TALK BACK TO YOURSELF

Back in the mid-1930s, as a school age lad, I found myself in quite a dilemma. I couldn't make up my mind whether I wanted to grow up to be another John Garfield of the movies, or Walter Winchell, writing a daily gossip column read by millions. Briefly, I considered a future as a "Human Fly," for such a daredevil gentleman had come to my home town and climbed the outside of the 12 story Hotel Vicksburg without ropes and with much acclaim. I finally settled for a career as Andre Baruch, a radio announcer.

Mr. Baruch was on the air often, and most noticeably as the exciting voice on YOUR HIT PARADE, a big, brassy, weekly show that surveyed phonograph record and sheet music sales all over the nation and presented the top tunes every Saturday night.

Radio's magic spell was everywhere, and I spent many evenings in my little room listening intently to WLW Cincinnati, WWL New Orleans and KMOX St. Louis on my primitive Atwater-Kent.

There was Baby Snooks, solely trying the patience of her long-suffering Daddy...Joe Penner asking "Wanna buy a duck?"...Major Bowes clanging his gong on the AMATEUR HOUR and brushing off starts of tomorrow with, "All right, all right!"...Eddie Cantor...Burns and Allen...Stoopnagle and Budd...President Roosevelt's FIRESIDE CHATS...Dance bands form the Black Hawk in Chicago...Amos and Andy.

Although program formats differed vastly in content and the big stars appeared and disappeared in quarter-hour to one-hour periods, I realized that an announcer was ever present.

He was the man who introduced the comedians and the music; he was the renowned authority on events of the day; he sat next to me at the championship fights and the World Series games. Yes, the announcer was the link between me and the outside world. And of all the voices on the air, I liked Andre Baruch's best.

Once I had chosen carefully my future career, I set out to do something about it. I managed to wheedle from my grandfather a prehistoric handcrank phonograph and a stack of old recordings that had gathered dust in the attic for several decades. This ancient equipment, together with some stimulating advertisements scissored from current magazines of that day and an aged Big Ben alarm clock, became my practice "radio station." For hours and hours, day after day, I introduced such doubtful song hits of the hour as "In the Gloaming" and "Bringing In The Sheaves," playing them over and over again. Interspersed with the music I extolled the virtues of Dr. Gragg's Foot Power, Mix-O-Mint Julep Mixer, Packard automobiles and other popular products.

Certainly, I was on the right track toward beginning a radio career, by practicing alone in a quiet room. Yet I worked under a serious handicap: I COULD NOT HEAR WHAT I SOUNDED LIKE! Thanks to electronic advancements through the years, a young fellow nowadqys who desires to go on the air as a professional does not face this problem. **A tape recorder solves it.** You can read a piece of copy into a machine; then instantly replay the tape.

If I were setting out to become an announcer these days, I would not do things much differently than I did in 1934, except to somehow acquire a tape recorder. I would keep practicing and listening to myself with a critical ear. I would put tapes recorded today aside, and listen to them the next week after they had a chance to grow "cold."

Above all, I would continue to practice with the tape recorder there to help me. I would not shout or raise my voice, but speak naturally with the microphone approximately eight inches from my mouth and a little to the side.

Daily I would schedule a specified hour for my "program," just as if I had a regular air shift at the local station. Afterwards, I would play back my show and criticize it. On my next practice program, I would attempt to perform better.

A television commercial produced for the telephone company demonstrates how, as a part of their training, beginning operators utilize tape machines to best learn how to inject friendliness and warmth into "May I help you?", "Number, please?" and so on. The voice, after all, is a "weather bureau" for the disposition. By its sound, we can tell whether to look for sunshine or warm weather, fog, storms, or freezing temperatures.

An investment of around $50 in a practical tape recorder can easily pay you back a thousand-fold in securing your toehold in broadcasting.

Not long ago I brought home a record album, "Themes Like Old Times," on which is transcribed 90 actual introductions of old radio shows just as they were once presented on the air. My wife and I listened with nostalgia to JACK ARMSTRONG—THE ALL-AMERICAN BOY, THE SHADOW, YOUNG WIDDER BROWN and dozens of show intros from the yesteryear.

Suddenly, out of the stereo speakers, burst the resonant introduction of THE STORY OF MYRT AND MARGE, a daytime serial.

"Why, honey" my wife exclaimed. "That announcer sounds a lot like you. Not his voice, exactly, but the way he phrases his sentences."

I smiled. The announcer she referred to was—you guessed it!—Andre Baruch. Without exactly planning to, I imagine that as an announcer-in-training I had copied his style.

GIVE YOURSELF A KICK
IN THE SEAT OF THE CAN'Ts

Nothing in the world can take the place of persistence in your desire to become a broadcast personality. **And persistence means daily practice.** Instead of doing your announcing homework everyday, it is much easier to watch TV, to go bowling, or bar hopping. It's much easier to put off today's self-help sessions until tomorrow.

During my early years as a kid hoping to get on radio, I was frequently tempted to put aside my well-worn phonograph records and dog-eared practice ads and take off for the movies and a chocolate soda with the good-looking blonde who lived next door. Somehow I forced myself to keep at my chores. In retrospect, I'm delighted that I made the decision. (Besides, the neat blonde married the high school football hero and soon got fat.)

You will find that persistence can be developed like any other habit. Few of us are born with the willpower to make ourselves keep at a task. This is a trait that we acquire, like learning to enjoy the opera or the eating of snails.

First of all, you must really want to get a job in broadcasting. Then, assuming that you are in dead earnest about starting, **your first step is to convince yourself that daily practice truly does pay off!** Once you get the conviction that honest effort is never lost, you will be happy to practice your announcing as regularly as the hands of the clock come around. Really bright people know that it's smart to be industrious and the pay-off is proportionate to the brains and effort invested.

An example is in order. In Seattle one of the most successful broadcasters is Lee Hurley. This man is a genius at building automated radio station equipment, and, at those times when he elects to go on the air, his deep, mellow voice and sexy delivery flutter many a feminine heart.

Lee Hurley is living proof that with persistence one can succeed. Lee first decided to be a broadcaster at the age of 15. From technical books, he taught himself how to build a radio station, which he accomplished within a few months and naively put on the air with only 25 watts of power and without approval of the Federal Communications Commission. The Commission, of course, immediately shut down the young man's operation.

Still fired with ambition, Lee next applied for an announcing position at a Yakima, Washington, station. He was told that to qualify for employment he had to be at least 18 years of age, that he had to have a first-class FCC engineering license.

For more than two years, Lee Hurley worked diligently towards his goal, rehearsing program presentations in his now silent little radio station and studying for his first-class ticket. On his 18th birthday, Lee came back to the Yakima outlet with his license in his pocket. They hired him. With enthusiasm, miracles are possible and practice is your best instructor.

A car will not start without a spark. And then after you have the spark, you must keep the motor running and put the car in gear to get anywhere.

ON TOP OF THE ACTION

Every phase of show business has its "bible." Movie people read **Variety** and the **Hollywood Reporter.** Advertising men, **Advertising Age** and **Printer's Ink.** Nightclub performers, carnival folk and recording artists, **Billboard.**

In radio and television, our bible is BROADCASTING, which is published weekly and sold by subscription only. There is also a BROADCASTING YEARBOOK published annually in which you will find the complete listing of all radio, TV and cable companies in the United States, together with the names of key personnel at each outlet. BROADCASTING publishes all the latest radio-TV information—new stations going on the air, stations changing ownership, personnel and formats, new ideas in programming, promotion and equipment, rulings by the Federal Communications Commission, and, most importantly, "Help Wanted" ads. You can get your name in the hat for a good position if you respond quickly to a classified in the magazine. There are many pages of opportunities from stations across the country. In one Arizona station, the manager keeps the magazine locked in a safe so his staff members can't read the want ads and move on to greener fields. If you want to subscribe, write to:

BROADCASTING PUBLICATIONS, INC.
1735 DeSales Street NW
Washington, D.C. 20036

I once knew a broadcaster who read the obituary column first of all, reasoning that whenever someone died there was an immediate job opening. Actually, he once got a $20,000-a-year position that way. He is now vice president for one of the TV networks. Perhaps he replaced a deceased executive in that slot by reading an obit.

WHAT ABOUT BROADCAST SCHOOLS?

The last time I looked there were hundreds of universities, colleges and schools (both public and private) in the United States offering radio-TV courses to thousands of students. Of these schools there are three types:

1. Education in classrooms where courses deal primarily with the principles and theory of communication and mass media research, and where little, if any, time is spent on the daily routine of station operation.

2. Education in broadcast departments where the main emphasis is on the creative aspects. That is, students learn how to develop ideas for programs, how to write continuity and shows, how to handle studio hardware, and how to announce.

3. Education by mail through correspondence courses, which enables the student to stuff himself with knowledge at the going rate of so-much-per-ounce, as determined by the U.S. Postal Service.

Will it do you any good to study radio and television at a college or to take a correspondence course? Yes, it will!

Everything helps. Approximately 70 percent of broadcast school graduates eventually find jobs in the industry, and many of the remaining 30 percent turn up in closely related work. Almost all professionals will advise you to take some type of training. The bywords are:

Study

Train

Improve

In picking a school of broadcasting, look for one with the best possible faculty—instructors who have professional backgrounds in the field as well as solid academic qualifications. Remember, a school is only as good as its teachers, so check credentials carefully to make sure the people you study under are **practical** broadcasters.

Obviously, you have to make your own investigation of the schools that are accessible to you, and the following listing of universities, colleges and schools offering broadcasting instruction will be helpful:

ALABAMA

University of Alabama
Tuscaloosa

Auburn University
Auburn

Alabama A & M College
Normal

Gadsden State Junior College
Gadsden

Jefferson State Junior College
Birmingham

ARIZONA

Arizona State University
Tempe

University of Arizona
Tucson

Phoenix College
Phoenix

ARKANSAS

Arkansas State College
Jonesboro

John Brown University
Siloam Springs

CALIFORNIA

University of California
at Los Angeles
Los Angeles

California State College
Fullerton & Los Angeles

Chico State College
Chico

Columbia College
Los Angeles

Fresno State College
Fresno

Humboldt State College
Arcata

Long Beach State College
Long Beach

Sacramento State College
Sacramento

San Diego State College
San Diego

San Fernando Valley State College
Northridge

San Francisco State College
San Francisco

San Jose State College
San Jose

University of Southern California
Los Angeles

Stanford University
Palo Alto

Bakersfield College
Bakersfield

Don Martin School of Radio &
Television, Arts & Sciences

El Camino College
El Camino

Foothill College
Los Altos

Fresno City College
Fresno

Heald College of Engineering
San Francisco

Pasadena City College
Pasadena

Pasadena Playhouse College
Pasadena

San Diego Mesa College
San Diego

City College of San Francisco
San Francisco

San Mateo College
San Mateo

COLORADO

Colorado State University
Ft. Collins

University of Colorado
Boulder

University of Denver
Denver

DISTRICT OF COLUMBIA

American University
Washington, D.C.

Capital Radio and Engineering
Institute
Washington, D.C.

Marjory Webster Junior College
Washington, D.C.

FLORIDA

Florida State University
Tallahassee

Florida Technological
University
Orlando

Pensacola Junior College
Pensacola

St. Petersburg Junior College
St. Petersburg

University of Florida
Gainesville

University of Miami
Coral Gables

University of West Florida
Pensacola

GEORGIA

University of Georgia
Athens

HAWAII

University of Hawaii
Honolulu

IDAHO

Idaho State University
Pocatello

University of Idaho
Moscow

Ricks College
Rexburg

ILLINOIS

Columbia College
Chicago

39

Eastern Illinois University
Charleston

Illinois State University
Normal

University of Illinois
Chicago

Northern Illinois University
DeKalb

Northwestern University
Evanston

University of Southern Illinois
Carbondale

Belleville Junior College
Belleville

Black Hawk College
Moline

Joliet Junior College
Joliet

Lake Land College
Mattoon

INDIANA

Ball State Teachers College
Muncie

Butler University
Indianapolis

Indiana State University
Terre Haute

Indiana Technical College
Fort Wayne

Indiana University
Bloomington

Purdue University
West Lafayette

Valparaiso Technical Institute
Valparaiso

IOWA

Drake University
Des Moines

Ellsworth Junior College
Iowa Falls

Iowa Central Community College
Fort Dodge

Iowa State University of Science
and Technology
Ames

University of Iowa
Iowa City

KANSAS

Barton County Community
Junior College
Great Bend

Hutchinson Junior College
Hutchinson

Kansas State University
Manhattan

Kansas University
Lawrence

University of Wichita
Wichita

KENTUCKY

University of Kentucky
Lexington

LOUISIANA

Louisiana Polytechnic
Institute
Ruston

Louisiana State University
Baton Rouge

Loyola University
New Orleans

University of Southwestern
Louisiana

MAINE

University of Maine
Orono

MARYLAND

University of Maryland
College Park

Baltimore Junior College
Baltimore

Community College of Baltimore
Baltimore

Hartford Junior College
Bel Air

MASSACHUSETTS

Boston University
Boston

Emerson College
Boston

University of Massachusetts
Amherst

MICHIGAN

Central Michigan University
Mount Pleasant

Ferris State College
Big Rapids

Kalamazoo Valley Community College
Kalamazoo

University of Detroit
Detroit

Michigan State University
East Lansing

University of Michigan
Ann Arbor

Wayne State University
Detroit

Western Michigan University
Kalamazoo

MINNESOTA

University of Minnesota
St. Paul

MISSISSIPPI

Jones County Junior College
Ellisville

Mississippi Gulf Coast Junior College
Handsboro Station

University of Mississippi
University

University of Southern Mississippi
Hattiesburg

MISSOURI

Central Missouri State College
Warrensburg

Central Technical Institute
Kansas City

Lindenwood College for Women
St. Charles

Stephens College
Columbia

University of Missouri
Columbia

Washington University
St. Louis

MONTANA

Montana State University
Bozeman

University of Montana
Missoula

NEBRASKA

Creighton University
Omaha

University of Nebraska
Lincoln

Wayne State College
Wayne

NEVADA

University of Nevada
Reno

NEW JERSEY

Seton Hall University
South Orange

Centenary College for Women
Hackettstown

NEW MEXICO

New Mexico State University
Las Cruces

41

University of New Mexico
Albuquerque

NEW YORK

Adelphi University
Garden City

Brooklyn College
Brooklyn

Columbia University
New York City

Cornell University
Ithaca

Fordham University
Hempstead

Ithaca College
Ithaca

University of New York
New York City

Queens College
New York City

State University College
at Genesco
Genesco

Syracuse University
Syracuse

Nassau Community College
Garden City

New York Institute of Technology
New York City

RCA Institute
New York City

Suffolk County Community College
Selden

NORTH CAROLINA

Lenoir Community College
Kinston

University of North Carolina
Chapel Hill

NORTH DAKOTA

University of North Dakota
Grand Forks

OHIO

Ashland College
Ashland

Bowling Green State University
Bowling Green

University of Cincinnati
Cincinnati

University of Dayton
Dayton

Kent State University
Kent

Miami University
Oxford

Ohio State University
Columbus

Ohio University
Athens

Western Reserve University
Cleveland

Xavier University
Cincinnati

OKLAHOMA

Oklahoma Military Academy
Claremore

Oklahoma State University
Stillwater

University of Oklahoma
Norman

University of Tulsa
Tulsa

OREGON

Blue Mountain Community College
Pendleton

Lane Community College
Eugene

Mt. Hood Community College
Gresham

Oregon State University
Corvallis

Portland Community College
Portland

University of Oregon
Eugene

Pacific University
Forest Grove

PENNSYLVANIA

Duquesne University
Pittsburgh

Penn State University
University Park

Point Park Clark
Pittsburgh

Temple University
Philadelphia

Penn Hall Junior College
Chambersburg

Williamsport Area Community College
Williamsport

SOUTH CAROLINA

Bob Jones University
Greenville

Tri County Technical Center
Pendleton

University of South Carolina
Columbia

SOUTH DAKOTA

Mitchell Area Voc-Tech School
Mitchell

South Dakota State University
Brookings

South Dakota State Teachers College
Springfield

University of South Dakota
Vermillion

TENNESSEE

East Tennessee State College
Johnson City

Jackson State Community College
Jackson

Memphis State University
Memphis

University of Tennessee
Knoxville

TEXAS

Abilene Christian College
Abilene

Amarillo College
Amarillo

Baylor University
Waco

Central Texas College
Killeen

East Texas State University
Commerce

Grayson County College
Denison

Odessa College
Odessa

North Texas State University
Denton

Southern Methodist University
Dallas

San Antonio College
San Antonio

Texas Christian University
Ft. Worth

Texas Technical College
Lubbock

University of Houston
Houston

University of Texas
Austin

University of Texas at El Paso
El Paso

Texas Women's University
Denton

UTAH

Brigham Young University
Salt Lake City

43

Utah State University
Logan

University of Utah
Salt Lake City

Webster State College
Ogden

VIRGINIA

Hampton Institute
Hampton

Stratford College
Danville

University of Virginia
Charlottesville

College of William & Mary
Williamsburg

WASHINGTON

Eastern Washington State College
Chenney

Gonzaga University
Spokane

Washington State University
Pullman

University of Washington
Seattle

Spokane Community College
Spokane

WEST VIRGINIA

Marshall University
Huntington

West Virginia University
Morgantown

WISCONSIN

Marquette University
Milwaukee

Milwaukee Area Tech College
Milwaukee

Milwaukee Institute of Technology
Milwaukee

Milwaukee School of Engineering
Milwaukee

Stout State College
Menomonie

University of Wisconsin
Madison

University of Wisconsin at
Milwaukee
Milwaukee

WYOMING

Central Wyoming College
Riverton

Northwest Community College
Powell

University of Wyoming
Laramie

For private residence schools of broadcasting and for correspondence training, I suggest that you check the classified pages in trade magazines. You will find them advertised in the "Instructions" section. If you do decide to train in an accredited broadcast workshop or by mail, give it that "old college try."

Once you commit yourself to a certain curriculum of study, stick with it for at least six months. Then judge yourself fairly. Do you perform better than when you started the course? If the answer is no, then you are wasting your time and money either because of (a) poor instruction, or (b) because you are trying to improve a talent that doesn't exist. Then either find another school, or enter another field of endeavor. To put it flatly: If you remain at the foot of the class, perhaps you should become a chiropodist.

TERMS OF THE TRADE

AM—Amplitude modulation.
ACROSS THE BOARD—A program which is broadcast Monday through Friday. (You would say, for example, that "The Today Show" is telecast by NBC-TV across the board.)
AD LIB—An extemporaneous remark or action not included in the script.
AFTRA—American Federation of Television and Radio Artists, labor union for many broadcasters.
AUDIO—An electrical signal produced by a microphone, turntable, etc., representing audible sound.
BACK TO BACK—Consecutive programs or commercials.
BG—Background.
BLUE GAG—Off-color material.
BOARD—Audio console in the control room.
BOMB OUT—As to "lay an egg."
BOOM—An overhead microphone on a movable arm.
BRIDGE—A link between two scenes, usually music or narration.
BUSY—Too complicated.
CALL LETTERS—The letters by which a station is identified. (WIND, Chicago and KOOL, Phoenix are two of the best.)
CANNED—Anything that is recorded.
CANS—Head phones.
CART—A tape cartridge.
CATV—Cable television.
CLAMBAKE—A loused-up show.
CLOSED CIRCUIT—Private program feed.
COLD—To read copy before looking it over, or to start a program without music or introduction.
COMMERCIAL—An advertisement in which we are offered the opportunity of a lifetime every few minutes.
CONTINUITY—Script or other material read by an announcer.
CONTROL ROOM—A sound-proof room, separate from the studio where an engineer controls volume and mixes sound.
COPY—Continuity or script.
CREDITS—The acknowledgement of performers, writers, director, producer and others connected with a program.

CU—Close up.

CUE—A signal to start or stop any element of a broadcast.

CUE CARD—A large cardboard with copy on it for shaky memories.

CUSHION—Extra material near the end of a program that may be used or eliminated, depending on the time factor.

CUT—Abrupt termination of dialogue or action.

DEAD AIR—A bad moment when nothing is happening.

DEAD MIKE—A microphone not in use.

DISC JOCKEY—A radio announcer who plays records.

DOLLY—Movable platform upon which a TV camera is mounted.

DRY RUN—A rehearsal.

DUB—To transfer material from one tape to another.

ECHO CHAMBER—A compartment that produces an echo effect.

ET—Electrical Transcription. A special broadcasting recording.

ETV—Education Television.

FM—Frequency Modulation.

FADE—To diminish or increase the audio or video.

FCC—The Federal Communications Commission, generally referred to affectionately by broadcasters as "THE Commission."

FEED—To transmit to another station.

FEEDBACK—A disturbing condition brought about by a return of amplifier's output to its input. (Also, used in reference to trouble with a sponsor.)

FLUFF—An unconscious mistake, usually in dialogue. (For example, an announcer on KNX, Hollywood, in plugging a baker's slogan, "The Best In Bread" fluffed by urging his listeners to enjoy the "Breast In Bed.")

FLIGHT—A schedule of spot announcements.

FORMAT—Program elements, arranged in an established pattern. (You would say, for instance, that WMAQ, Chicago, has a "good music format.")

FROM HUNGER—A bad performance, or bad copy.

45s—Recordings that play at 45 revolutions per minute.

FREE LANCE—A performer who is not a member of the staff.

FROM THE TOP—To start from the beginning of a broadcast or script.

GAG—A comedy device.

GAIN—Degree of audio volume.

GIMMICK—A new approach or angle for a program or commercial.

HOT MIKE—A microphone that is turned on.
KILL—To do away with.
LAY AN EGG—As to "Bomb Out," to meet with something less than success.
LEVEL—Amount of volume being transmitted.
LIVE—Happening now, and not taped or recorded.
LOCAL—A program heard over only one station, as opposed to a regional or national network.
LOG—A station's complete schedule. A written record required by FCC regulations.
MASTER CONTROL—The heart of the engineering part of a broadcasting station.
MC—Program Host.
MIKE—The microphone.
MONITOR—A video screen for viewing or listening to a program.
MUDDY—Audio that fails to carry sharp, lifelike qualities.
NARRATION—Descriptive speech.
NETWORK—Two or more stations united by telephone wires or microwave for the simultaneous transmission of programs.
OFF MIKE—Not speaking directly into the microphone.
ON THE BUTTON—On time!
ON THE CUFF—Performing for free.
ON THE NOSE—Running exactly on time.
ONE SHOT—A program that is not part of a series.
PD—A station's program director.
PLAYBACK—Playing a tape or recording.
PLATTER—A record.
PLUG—A commercial announcement.
PSA—Public Service Announcement.
REMOTE—A broadcast that originates outside of the studios.
RIDE GAIN—To regulate the volume of transmitted sound.
ROS—Run of Schedule. Spot announcements sprinkled throughout the broadcast day.
RUN THROUGH—Rehearsing continuity.
SAUSAGE—A TV commercial that is hastily produced.
SCRIPT—Written material to be read on the air.
SEGUE—Transition from one musical number to another without a break.
SIGNATURE—A theme that identifies one program from another.

47

SIMULCAST—A program broadcast on both radio and TV simultaneously, such as an address to the nation by the President of the United States.

SOAP OPERA—A daytime serial.

SPOT ANNOUNCEMENT—A commercial.

SPONSOR—An advertiser who, according to one wit, makes a program possible and impossible at the same time.

STAND BY—A reminder to clear your throat and get ready.

STATION BREAK—A pause in programming to allow an announcer to identify his station.

STRETCH—A cue used to slow down a performer.

SUSTAINING—Unsponsored.

TAKE IT AWAY—A cue to begin.

TALK BACK—A communications system permitting persons in the studio to speak with control room personnel.

TECH—A studio technician.

TELEPROMPTER—A mechanical device much like a player piano roll from which a performer reads his lines.

TIGHT—A program or commercial in which there is too much material for the time allotted.

TURKEY—A flop.

TURNTABLE—A device for playing records and transscriptions.

VTR—Video Tape Recording. A mechanical reproduction of both audio and video on tape.

VOICE OVER—Narration over the film or video tape.

WOW—Distortion at the beginning of a record or during its playing, caused by varying turntable speed.

Getting Your Start

PART 2

HELP WANTED—EVERYWHERE!

People who are out to make their marks in the broadcasting world usually start off with a question mark: Where do I look for a job?

In answering, I'd say **right around the corner!** Almost every town has at least one radio station, whereas larger cities boast multiple stations, both AM and FM, as well as television studios. In addition, there are cable TV companies originating local programs, advertising agencies hiring talent for commercials, film producers and program packagers putting men and women on the air regularly.

A list of names and addresses printed here would not match the permanency of the other material in this book, for they are constantly changing. Therefore, our index of the marketplace must be a general breakdown of the various media and the job categories in each one.

To start with, no matter where you live—in hamlet, burg or metropolis—a search of the Yellow Pages in your local telephone directory will turn up prospective employers. What's more, I will wager that at every broadcasting facility there is at least one job opening at this very moment, or soon to be one.

Announcers, especially, are a restless breed of cat, always on the move for a better paying job in a bigger market, or merely for a change in climate. They are ready to hit the road for greater things at the flick of a microphone switch. You will learn that it is truer in broadcasting than any other walk of life that the grass always seems greener on the other side of the hill.

In the years that I managed radio stations in small towns, I cannot recall a single week in which I was not hard pressed for a replacement for some staff member. The disc jockey spinning 45s on the midnight to dawn show had received a long-distance call from a program director somewhere offering him a daytime shift, and he was leaving immediately.

49

The woman commentator was either pregnant; she was quitting work to stay home with the kids; or her husband was being transferred by his company to Flushing. If single, she was either getting married, or several months along with child, or both.

The newsman was leaving us to try his luck in Hollywood or New York.

Our afternoon deejay had got the chief of police's daughter in trouble and had skipped town by moonlight; or he had been busted for smoking pot or for passing a bad check.

There have been days when I might have been tempted to employ Quasimodo, the tongue-tied Hunchback of Notre Dame, if that poor soul had lurched in seeking employment. That's how desperate I sometimes got seeking staff replacements.

In radio and TV, personnel seem to be coming and going constantly. Perhaps this is one reason why broadcasting makes for such an exciting and fun business. To say the least, it is never static.

Many times over the years when that gypsy feeling has come over me, I have thought fleetingly of piling my loved ones in the family heap and rolling across the United States, working for a month or two in Phoenix, then on to New Orleans, Atlanta, Miami, or wherever. The possibility that jobs would not be immediately available has never crossed my mind. That's how confident I am that there is always a position open for an announcer, time salesman or copywriter.

As an example, an announcer whom I knew quite well at a California radio station got fed up with what he felt was unreasonable harrassment from his ex-wife. One day, on impulse, Jack, as I will call him, quit his job and set out for Florida in his battered Ford, together with his new bride and her two children from a previous marriage. Jack had no job leads whatsoever, and too little money—just a couple of hundred dollars and a gasoline credit card.

Upon arriving in the Sunshine State, he spent a few days looking over the area, driving down one coast and up the other, deciding on which Florida city most appealed to his family. They decided on West Palm Beach. Down to his last few greenbacks, Jack applied for work at the leading station there—**and got a job!** What's more, it paid twice the salary he had earned on the West Coast.

Jack's adventure took place six years ago, and the last time I looked he was going great guns as one of the highest-paid personalities in Florida. He had made the down payment on a new home, bought a boat with a flying bridge, and had money in his savings account.

Next we come to Don, who must be the world's original itinerant disc jockey. Don is a broadcaster who, for the past 15 years, has moved from radio station to station across the country, changing jobs on the average of once every three months—and never at the same station twice.

His winters are spent in Florida, Arizona or California; his summers in Washington, Oregon or Idaho; and during the spring and fall seasons, he works in New York, Illinois, Louisiana or Texas. Don loves to travel and meet new people (especially women) and, in broadcasting, he has found the right field to satisfy his wanderlust.

As a clinical experiment while writing this book, I placed a so-called blind RADIO-SITUATIONS WANTED ad in Broadcasting Magazine at their 30 cents per word rate. It read as follows:

"Announcer prefers Southern State. Best offer. Box C----"

Please note that I **did not** claim to be an **experienced announcer**. Yet I received a total of 18 replies from radio stations, ranging in power from 250 to 10,000 watts, mentioning starting salaries from $90 to $165 weekly. Most asked for audition tapes; some did not.

Once you make up your mind to pursue a radio-TV career, and you have carried out your practice exercises diligently, your first break lies just beyond that next bend in the road.

THINK SMALL

Without doubt, the best training ground for a broadcaster is the small, local radio station. Since those early days when Mr. Marconi's radio waves first joined balloons and smoke rings in the air, small towns have been the historic province of broadcasting apprenticeship. In the so-called "boondocks," a performer gains experience and poise. Backwaters are witness to evolving styles.

As a staff member of a little station somewhere, you will be called upon to disc jockey musical programs, deliver newscasts, call play-by-play sports, interview guests, go out and collect news stories (and possibly money owed to the station) and voice a mixed bag of commercials. In all likelihood, you will be required to write some commercial announcements and dream up station promotions. You may even be invited to sell radio time to local merchants for a few extra bucks. Fledgling announcers at "coffee pot" stations, as they are sometimes known, are usually required to do everything which has to do with broadcasting. And that's good!

By starting at a small station, you get a chance to develop all of your abilities to a point where you are fully prepared to go on to any position in a larger city at a higher pay. You gain insight and experience into every avenue of the broadcasting profession—something you cannot hope to accomplish if you should take your first job at a hugely staffed 50,000-watt powerhouse in a metropolitan center. At the latter your duties are **too specialized** for you to acquire a well-rounded education as a broadcaster.

Many of today's top stars, station owners and network executives started out by thinking small and by learning their trade at low pay at local stations. For example, Bob Crane, best known for his starring role in HOGAN'S HEROES, first broke into radio in Connecticut as a disc jockey where he learned to handle anything. Bob became an accomplished announcer, newsman, MC and actor. In due time CBS Radio hired him. Six months after that breakthrough he struck big-time paydirt.

Tom Campbell, a cool jock in his 30s, and one of the San Francisco Bay Area's most popular mikemen, now earns a reported $100,000 a year or more. But Tom makes it clear that the journey to the top wasn't easy. He points out how he spent years learning his trade. Small stations and even smaller paychecks throughout the South was the price he paid for getting where he is today.

BATMAN Adam West entered radio at a station in San Luis Obispo, CA. He later joined with an old Army buddy and they became funny as a pair of ad-libbing disc jockeys, hosting a conversation-and-variety program in Honolulu called the KINI POPO SHOW. From Hawaii Adam West crashed into Hollywood films.

The list of superstars who served their apprenticeship in small-time radio seems limitless and I could not begin to mention them all in these paragraphs—Art Linkletter, Jack Paar, Arthur Godfrey, Paul Harvey, Steve Allen, Phyllis Diller, Rod McKuen, and so many more. In the beginning they all thought small.

In most fields today, every job seems to call for a college degree. Not so in broadcasting. The experience you acquire, your familiarity with the structure and the vernacular of the industry, and the understanding of its demands will be invaluable to you upon taking advantage of your first opportunity.

Just imagine your future in broadcasting as a baseball diamond, with a 250-watt station in the grass roots as first base. To score in the long run, you must start off with a hit and land on first base. As your hitting improves, you can try for a homerun. LITTLE STATIONS IN LITTLE BURGS ARE THE STEPPING STONES TO STARDOM.

MEN FOR ALL SEASONS

It pays to learn how to do all the jobs in a radio or TV station; not just what you like to do, but EVERYTHING! Versatility compensates far better than specialization. A person who can "wear any hat" in the media increases his job security and improves his chances for advancement to big-city stations and to networks.

The behind-the-mike story of Hugh Downs, the well-known NBC personality, is typical. Just out of high school, Hugh served his apprenticeship in the Ohio "boonies" as an announcer. As time went on, he learned a bit about engineering; he built special programs, wrote commercial copy and sold radio time. From the small grass-roots station, Hugh joined NBC's staff in Chicago and the network eventually brought him to New York after he had built a reputation in the Windy City.

Over the years I have shared the microphone with men and women whose versatility in and about a station guarantees them steady employment anytime, anywhere. Foremost in my mind of broadcasters competent of many things is Rudolph Von Tobel, a man who has never yearned for the big leagues of the industry, but a gentle fellow who prefers small town living and the challenge of working for smaller stations. I am confident that he would now be one of the best-known names in the business had ambition pointed him in that upward direction. The talented Mr. Von Tobel began his broadcasting career by wisely taking three important steps:

1. In the U.S. Naval Reserve, as a radio operator, he learned broadcast engineering.

2. At first, he **thought small** and located a low-paying position at a 250-watt station in Nevada, building up his knowledge of station operation piece by piece.

3. For professional purposes he shortened his name to "Rudy Von."

As the years passed, Rudy moved up the ladder to larger stations in more important markets, while purposely not rattling any doors in Los Angeles or New York.

"Too dog-eat-dog," he commented. Today he is so knowledgeable of broadcasting that a station owner can confidently hand him the keys to the place and take off for an around-the-world cruise, satisfied that Rudy will keep it running successfully.

First and foremost, Rudy is an announcer with a deep, rich baritone voice. He can read commercials in several styles of delivery, announce any kind of sporting event from Little League softball to professional soccer, create musical shows, compile and read the news, cover special events and welcome on the air visiting dignitaries with a flair that would make a diplomat envious. He is an adroit ad libber.

If something breaks down in the control room or transmitter, Rudy can fix it. He knows the "nuts and bolts" of radio engineering and has a first class FCC license to back him up. To my knowledge, Rudy has failed only once to keep a transmitter humming. That was the interesting day that a large bullfrog somehow got trapped in ours and electrocuted itself while blowing the station off the air. Croaked its last, so to speak.

In a metropolitan station or at a network, Rudy's abilities thus far described would more than justify a fat paycheck. But in our 1000-watt hometown station in the Northwest, it was only for starters.

Once Rudy had completed his 9 AM to 3 PM air shift, announcing, keeping the logs and mothering the equipment, he slipped on a jacket, knotted his tie neatly, and began making his rounds as a time salesman. He could always be depended upon to add to the station coffers with advertising revenue from a variety of merchants. In came contracts of his sales to dance halls, coin-operated laundries, skating rinks, appliance stores and other places of business. Next, he wrote their commercial copy and taped these spots for broadcast.

Rudy now played another equally important role—as a bill collector. Sponsors who did not pay their advertising tabs after 60 days were relentlessly hounded by Rudy until they did.

Then, in the evening, after a long day at a hot mike exalting the praises of demanding advertisers, pecking at a typewriter and pounding the downtown bricks scrounging for business or accounts receivable, Rudy could often be located at the station taping special shows for broadcast on Sunday, his day off. And incidentally, if the janitor failed to show up, Rudy usually volunteered to push a broom or carry out the day's barrel of trash. Wherever Rudy Von is hanging his magic hat these days, lucky is his employer, for he has an all-around radio man on the payroll!

I would hope, dear reader, that your mind is now made up to become another Rudolph Von Tobel.

Please keep in mind that the more you understand of broadcasting techniques, the better this will serve you in the years to come. It is important to learn the total routine of station operation, the handling of turntables, tape players and control equipment, program traffic, sales, billing and promotion.

"All right," you say, "after six months or a year on a local station, where do I go from here?"

It depends on what you have learned in that period of time. You may find out that while you have a certain amount of talent to knock 'em dead on a hometown station, you cannot measure up against the increasingly stiffer competition you will meet in the big time. In this case, you can take off in a new direction.

Announcing is just a first step of getting into the industry. In broadcasting today are countless executives, writers and directors who began at the microphone. As an executive, or a director, you will be better qualified to evaluate and guide the work of your announcers if you understand them and their work. As a writer, you will be better programmed to create sharper copy for oral presentation.

By becoming a "man of all seasons" you also become a man of all reasons to be hired and promoted, step by step, to the top of the profession.

In these pages I have referred exclusively to men. Yet we must consider the ladies, too. The industry is lucky to employ innumerable women who are female counterparts of Rudy Von. Gals of many broadcast skills.

Barbara Mowbray comes instantly to mind. She is a terrific and prolific person. Barbara and her husband, John, a veteran broadcaster, now own and operate KQIN, a daytimer in Burien, Washington, a suburb of Seattle. As Girl Friday of the station, Barbara is responsible for a daily calendar of chores that staggers the imagination. Her duties include: receptionist, bookkeeper, traffic girl, promotion manager, music librarian, home economist, copywriter, sales woman, personnel director, and commentator. To this impressive list we must further add, "Etc., Etc."

Barbara, who got her early training in New York drama and later at radio stations in Central California, is a top professional and a valuable asset to any station. When not at KQIN, doing her things, Barbara cooks and sews for six children and raises prize-winning bulldogs as a hobby. As her husband has stated: "Without Barb, the damned station wouldn't go—or the Mowbray household, either!"

The moral of our case histories, as they apply to you, is simple: If you have talent and cannot use it, you have failed. If you have a talent and use only half of it, you have partly failed. If you have a talent and learn somehow to use all of it, you have succeeded and won a satisfaction that few men and women ever know.

As you plan your assault on the radio-TV media, think of the Marine Corps. In this elite outfit a recruit is taught to disassemble his rifle and reassemble it blindfolded. Put your broadcast job in this same perspective. Take steps to find out enough about the inside of a station so that you can keep it going all by yourself if you have to. **Analyzing the typical staff member's role in the radio-TV industry, it pays well to be a "know it all."**

OPEN MIKE

At this point let us set the record straight. Regardless of the amount of talent that you possess, the odds are many thousands to one that you are not going to start your career as a performer on a national radio or television network. This sort of thing rarely happens in today's competitive world.

You may be Frank McGee, Harry Reasoner and David Frost rolled into one. Or a feminine ball of ability with all the attributes of Mary Tyler Moore, Nanette Fabray and Betty White. Nevertheless, over the coming months, and probably years, you will find it necessary to prove your special genius to a great many skeptics and take your turns moving up the line.

Since we must take for granted that you will step into the broadcasting field in a small way, we will now put a typically local station under the microscope. We cover as many areas of potential employment as possible, giving concise job descriptions, which are general from market to market and station to station. To start with, you may have a few questions buzzing about in your head.

Possibly you are wondering why some station call letters begin with W and some begin with K? The answer lies in radio's infancy. During that era in the 20s, most radio stations were located in the Eastern United States and they were assigned W call letters.

As station towers began to sprout up across the country, the Mississippi River was made the dividing line—Ws in the East; Ks in the West. There are a few rare exceptions to this rule, because these were among the earliest stations and they adopted their call letters before the call-letter system was as rigidly controlled as it is today. KDKA, Pittsburgh, Pa., which went on the air November 2, 1920, is a good example of a K station in the East.

Most of the early broadcasting stations were allotted three letters—WOR, WGN, WSB, KFI, to name a few—but once the 3-letter combination became exhausted, it was necessary to add a fourth letter.

Considerable brain strain goes into the selection of call letters. It is not a catch-as-catch-can situation. Station owners apply sometimes for pronouncable combinations like WIND, KORN, KOLD and KING, or for zippy call letters that are easily remembered by the average listener. WHIZ, KICK, WHAM and KAYO illustrate the point. In some instances, the initials of station owners appear in call letters. KNBC, WCBS and WABC stand for National Broadcasting Company, Columbia Broadcasting System and American Broadcasting Company, respectively.

Applicants may chose any arrangement of four letters beginning with the appropriate W or K, provided they are not identical with the call letters of an existing station and are in good taste. Combinations such as KRUD and KRAP would not qualify.

While on the subject, we should mention that stations in Canada employ call letter combinations beginning with C and in Mexico with X. International agreement apportions the alphabet among nations.

Enter the Federal Communications Commission. **Just what is the FCC?** Surely you have heard of the life and death power that this government agency wields over broadcasting stations, but you should know a bit more about it.

To police the airways, and, to a certain extent, to police station owners, Congress established the FCC by enacting the Communications Act of 1934 as an independent federal agency composed of seven Commissioners appointed by the President.

Suppose you wish to operate a radio or television station? Without exception you must apply to the Commission for a license. As part of your application, it is mandatory to show evidence of moral and financial responsibility, of your ability to operate in the "public interest, convenience and necessity," and you have to demonstrate engineering evidence of available frequencies or channels. Upon satisfactory submission of such evidence to the FCC, a **limited** license may be granted to you. But, this license has to be renewed periodically, inasmuch as the FCC retains the continuing power to suspend or revoke it for various reasons.

All stations are required by the FCC to keep program and transmitter logs, minute-by-minute written records of everything broadcast. What's more, FCC inspectors make **surprise inspections**, trick-or-treat kind of visits that can send shudders down the spine of the operator on duty who has not kept his log up to date, or who is not keeping his broadcast

57

FCC Rules and Regulations require that transmitter readings be logged regularly. In this shot, the combo man on duty performs the necessary function.

signal within a specified pattern. Woe betide the station with too many violations! Penalties for broadcast station violations, depending upon their degree of seriousness, range from reprimands, fines up to $10,000 and short-term licenses, to outright denials of license renewals or the revocation of the licenses.

How does a broadcasting station bring in enough money to pay its bills and to return a profit to the owner? At first glance, it would seem that this question could be covered in a single sentence: Rates for commercial time are established by each station after a careful study of the approximate number of listeners who might be expected to hear the messages at the time of broadcasting, and this commercial time is sold to advertisers. However, there are vital points of difference between conditions that make a sale.

In general, sales departments are divided into three categories:

1. Local

2. Spot Sales

3. Network

On the local level, the size of the sales department is determined by the size and importance of the commercial station. Usually, there is a sales manager who oversees a staff of salesmen, plus sales secretaries, a traffic department and billing clerks.

Each salesman, knowledgeable of all the pertinent facts about his station (or so the sales manager hopes) calls on prospective advertisers and sells as much air time as possible.

After the advertiser has contracted for a certain number of spot announcements or time periods, his commercial copy is prepared by the program department, scheduled for broadcast by the traffic department, and presented on the air. Following this the advertiser is sent a bill, which hopefully he pays post haste.

As for spot sales, some national advertisers, through their advertising agencies, prefer to buy individual stations across the country with the purpose of marketing their products through selectivity. Therefore, almost all commercial radio and TV stations are represented by firms, called station "reps," in principal cities. Working on a commission basis of 15 percent these reps contact the ad agencies and sell commercial time for their hometown station clients.

Thirdly, there are the networks, which are organized to connect hundreds of stations together in order to make possible the broadcasting of a single program or spot announcement to the entire country at the same time. The networks serve the needs of those national advertisers, like the soaps, the patent medicines and the breakfast cereals, who

59

choose to reach a coast-to-coast audience on a web of stations, rather than by buying schedules on hundreds of stations individually. Networks are paid high rates by the advertisers for these commercials; in turn, the networks share part of this revenue with stations carrying the commercial messages.

To put station sales in a nutshell, **every time a station broadcasts a commercial, it is earning money directly from the advertiser or from the network.** The station in turn uses this revenue to pay for its electricity, its equipment, its news services, its insurance, its announcers, engineers, sales people, office help and other expenses. The owner of the station prays that after this outlay of cash there is something left for his own pocket.

Without being technical about it, what is the difference between AM and FM broadcasting? Unless you possess a great deal of engineering knowledge, an explanation of AM-FM methods of transmission is confusing.

Stations on the standard broadcast band are called AM stations and they broadcast on frequencies from 535 kHz to 1605 kHz (or kilocycles as it used to be called). The other type of radio stations are known as FM stations and they broadcast on frequencies from 88.1 to 107.9 MHz (or megacycles). To put it simply, AM stands for Amplitude Modulation, and, as of this date, most of the successful commercial stations are of this type.

Because there are so many more AM stations in the United States than FMs (over 4000 AMs as opposed to about half that number of FMs) the AMs enjoy larger audiences and greater financial rewards, in most cases. Homes and automobiles presently are equipped to receive AM primarily, whereas most receivers are not able to pick up FM broadcasts.

On the other hand, FM, which stands for Frequency Modulation, gives better reception because it is generally free of static and interference. Lightning, electrical appliances, elevators in the neighborhood and other such distrubances do not interfere with FM as they do with AM reception. Music presented on FM has exceptionally fine qualities, and this explains why so many FM stations specialize in classical and semi-classical selections played by large orchestras.

Will FM radio eventually replace AM as the carrier of broadcast music due to its clear fidelity? An increasing number of broadcasters think so. The biggest thing standing in the way right now is the fact that FM penetration is still not up to AM in the home audience, and it trails sadly in the automobile audience, radio's bread-and-butter market.

However, auto manufacturers, in Detroit and abroad, currently are including FM with AM in up to 30 percent of the radios installed. In another few years, it is anticipated that this will go to 50-50. Then higher. FM penetration in homes continues to grow dramatically, and FM's future is increasingly formidable.

Why do some radio stations broadcast full-time and others only part of the day? The FCC grants licenses to AM stations to operate according to the following time schedules:
1. UNLIMITED TIME permits a station to broadcast 24 hours a day if it so desires.
2. DAYTIME ONLY stations may operate solely between sunrise and sunset.
3. SHARING TIME and LIMITED TIME schedules permit stations to operate when other stations sharing their frequencies are off the air.

The radio station formats, how do they vary? Just as a fellow may prefer a redhead, rather than a blonde or brunette, so it is with radio audiences. Some people wish to hear songs by Tom Jones, while others desire the Latin rhythms of Big Boy Bolo and His Bongo Beat. Because people are different, radio formats must be different. For this reason, radio stations specialize in distinct types of programming to satisfy various segments of the listening public and to guarantee tune-ins.

National surveys in recent years show the largest single category to be what is known as contemporary music (rock 'n roll). Stations making use of this most popular format usually take their music selection from the Top 40, 50 or 60 tunes in the nation, based on record sales, and play these on the air over and over again much like a jukebox repeating itself continually. There are jingles, jokes and contests. News is usually presented in brief, terse segments.

Second in popularity come so-called middle-of-the-road stations, broadcasting a blend of hummable, whistleable tunes, both old and new, along with regularly scheduled newscasts, presented most often on the hour and half hour.

Thirdly, modern Country and Western stations are popular most everywhere, but especially in the South and Southwest. That "Nashville sound" has moved up steadily to "grab" a large segment of the public.

Behind these leaders come rhythm and blues, ethnic and religious stations, album, standard and classical stations. The all-news, all-talk broadcasters fit into still another niche. In some major cities they rank among the top leaders in popularity.

Ideally, each successful station, regardless of its program policy, tries to establish a definite sound that will appeal to the listener. After this is accomplished, **the station looks for the right air personality who will work with the music (or other type of format) and not compete with it for attention.** That's where you come in.

Is an announcer in a typical radio station setup required to hold an FCC engineering license? Yes, in most small stations. A disc jockey who works in a control room has to have at least a third class radiotelephone operator's permit in accordance with FCC Rules and Regulations.

To get such a license, a written examination is necessary. But, take heart! If you are a non-technical person, as I am, there is no cause to panic simply because something of a technical nature has unexpectedly been placed in your path. The examination is quite simple. Questions in the test concern your overall knowledge of station operation: Who keeps the station logs? How often should station identification be made? When should a program be announced as "recorded"? You won't have to know what a "framish" is, or how it activates a "do-dinger."

The following questions and answers typify the type of information that will appear in the examination, except they will be the multiple-choice type:

1. WHO MAY APPLY FOR AN FCC LICENSE?
    Commercial licenses are generally issued only to citizens of the United States.

2. WHAT TYPE OF COMMUNICATION HAS TOP PRIORITY IN THE MOBILE SERVICE?
    Distress calls.

3. SHOULD A TEST OF RADIOTELEPHONE EQUIPMENT BE MADE DAILY, WEEKLY OR ONCE A MONTH?
    Daily.

In preparing for this examination, you should write to the:

Superintendent of Documents
U.S. Printing Office
Washington, D.C. 20402

News cars, equipped with shortwave radio, help make it possible for news-oriented stations like KXRX to keep on top of what's happening in their marketing areas.

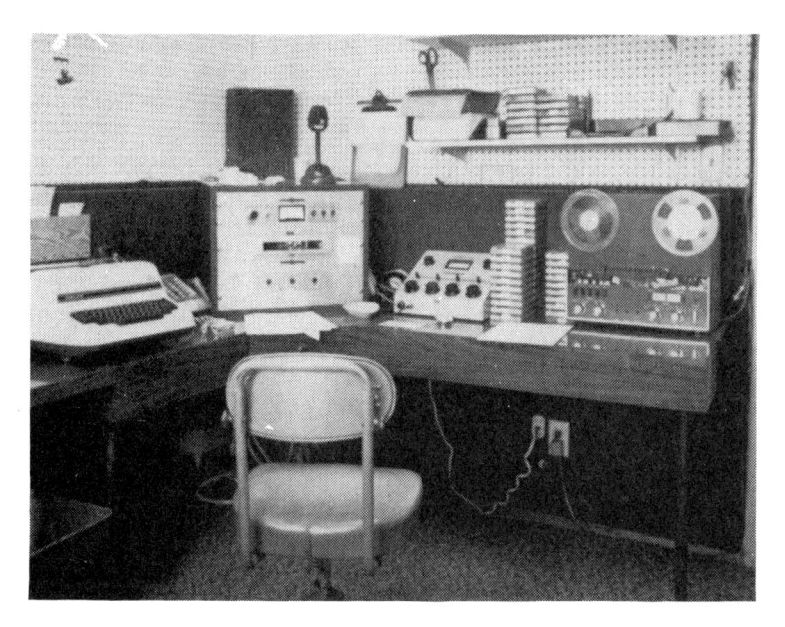

The KXRX newsroom.

63

Radio newsmen keep on top of the action in the KXRX newsroom.

and ask for FCC FIELD ENGINEERING BULLETIN NO. 4. This piece of government literature contains all the information that will ready you for the test.

You should also receive from the U.S. Printing Office a listing of the location of the Offices of the Federal Communications Commission where you may take your examination. FCC offices are located nationwide in major cities.

In my own case, I am stumped by most anything mechanical. My wife even sets my watch. Yet I have operated complicated radio station equipment as a licensed operator for many years. Once you get the hang of it, the routine becomes as easy as a chimpanzee peeling bananas. Take my word for it, if I can obtain a radiotelephone license, so can you!

What advantage does an announcer have by holding a first class license? The federal government clearly states that in specific cases an engineer with a first class license, or "ticket" as it is called, be on duty at a station during broadcast hours.

Engineers frequently handle announcing duties in small stations for salary-saving purposes, and they are known as "combination" or "combo" men. Lots of stations require that all operating personnel be combo. They like to have doubly qualified people in preference to those with single skills. Sometimes an announcer who has a ticket is stationed at the transmitter performing technical duties of a limited nature.

Obviously then, one of the very best ways of breaking into the radio field is by securing a first-class license. As an announcer holding such a ticket, you are immediately considered more qualified and usually, as a result, you are better paid. I strongly recommend that, as an announcer-in-training, you take steps to obtain such a ticket. Your opportunities will at least be doubled, and I would predict that your chances will

In the news booth a radio reporter prepares continuity for his upcoming newscast.

This announcer is cueing a tape for an upcoming program.

be increased a hundredfold in finding a starting point in the radio industry.

How does a person obtain a first-class radiotelephone license? There are three choices. You may study the subject matter by yourself from technical books; enroll in a course of home study that is designed for a beginner; or, if finances permit, enroll in an engineering school. The chief engineer of your local radio or TV station is a good man to talk to. He will set you straight on the best ways and means of obtaining your first phone.

TAB Books, Blue Ridge Summit, Pa. 17214, is the publisher of many books that can be especially helpful to you. For example, veteran broadcaster Hal Fisher, author of HOW TO BECOME A RADIO DISC JOCKEY, a TAB publication, devotes a special chapter titled "Obtaining Your FCC License" to the subject. A copy of this fine book would be most helpful to you in planning your career. There are also study guides with questions and answers.

What if I can't crack the FCC examination for a first class ticket because I'm not mechanically inclined? Fooey! I am acquainted with dozens of announcers holding first class licenses who are close to being "mechanical morons." Frankly, the pencil sharpener in the station office is about as far as they ever get in the line of operating a complicated mechanism, let alone working with highly sophisticated radio equipment.

These people are graduates of what I would call "quickie" or "shotgun courses" conducted at certain schools. In such operations, classes operate seven days and five nights a week from 9 AM until around midnight, broken only by time-outs for meals and short coffee breaks.

After six to nine weeks of having your head crammed with technical data, you can walk out of this course with enough knowledge to pass the FCC exam for your first ticket. You may forget most of what you were taught in a short time, but, in the meantime, you will have acquired your ticket to hang on the control room wall and satisfy the FCC. The classified advertising section of trade magazines, recommended earlier in this book as a source for employment, likewise give you a line on good schools to check with.

A TYPICAL STATION

Radio, to the average listener, signifies songs by Sinatra, music by the Tiajuana Brass or weekend football. It signifies

In the main control room, the disc jockey cues a record. Notice that he faces a newscaster in the adjoining studio through a glass panel. This arrangement makes for better communication between performers and tighter production.

foamy-voiced beer salesmen and serious voices praising headache and backache remedies and suppositories. It represents addresses by the President, election returns, news coverage and stock market quotations. Radio to the general public is music, information, opinion, contests and disc jockey quips. The average listener either loves or hates what he hears. He doesn't know exactly how it gets on the air. He merely thinks of radio as good or bad.

Behind the scenes, the broadcaster thinks of such things, too, but with a profound difference. He wishes his output of programming to be as good as he can make it with the limited staff available to him.

The primary function of any commercial station is to produce programming and to sell interruptions within it to advertisers. The mode of operation depends upon the location of the station, its signal strength and the money available to keep it going. Many daytime-only stations of 250 to 1000 watts operate 365 days a year with as few as five people who double as announcers, engineers, salesmen and bookkeepers.

For purpose of demonstration, we are going to create in the next pages a hypothetical 1000-watt full-time station. You and I will be its owners. We will call it Radio KZAM, and analyze how it is staffed, how it functions and how jobs in KZAM are obtained.

Because we will concentrate strictly on local live disc jockey programming, supported chiefly by local advertising, we must employ:

(a) A staff of salesmen to bring us business.

(b) A traffic department to process all sales orders.

(c) Program people to plan and produce programs.

(d) An accounting department to bill for advertising and collect monies.

(e) Engineers to keep us on the air.

We are putting KZAM in a city of 50,000 population, which will name Zamville, U.S.A. Of first importance to us is the building in which we are to locate our facilities. For convenience sake, we are erecting our structure at the transmitter site so that everything is handy.

Considerable planning has gone into the construction of the Radio KZAM studios, for these make up the heart of the

69

operation. First, there is the master control room, suitable in size and shape for the continuing activity that takes place inside it. Everything else in the station is built around it.

This sound-proofed control room accommodates the disc jockey on duty and is so constructed that where he is seated at the microphone the DJ has complete access to everything that he needs without getting out of his chair—the console, or control board, clock with sweep second hand, program and transmitter logs, 45 RPM records, 33 1/3 music albums, four turntables that play at any speed, tape racks full of tape cartridges, tape machines and commercial copy.

Separated from master control by a large glass window is a small announcing booth, from which news and sports are read and special announcements made.

Close at hand is still another small studio with a control board, several turntables and tape machines. In here announcements and special features are prepared for upcoming broadcasts. An intercom system for communication connects these studios.

The news printers (teletype machines continuously feeding material from Associated Press and United Press International wire services) are also nearby, situated in a special sound-proofed chamber that is easily accessible to all members of the KZAM staff.

The reception room, where station visitors are greeted pleasantly by our girl friday-receptionist, is just outside of these areas, yet isolated from the studios by thick panes of glass that keep out noises. Visitors outside can see and hear what is going on without interfering with any of the production activity.

Elsewhere in our highly functional building are offices housing the program department, engineering, the sales people, bookkeeper and station manager.

Radio KZAM is staffed with a total of 16 people, and each of them puts in about 40 hours a week. In addition we have some outside help. We retain an attorney in Zamville for any legal problems that arise locally. And we have legal representation in Washington, D.C. for the handling of licensing matters with the FCC. We pay a monthly retainer to an independent Certified Public Accountant (CPA) to keep us periodically advised of profits and losses.

Taking our cue from the most popular formats in the nation, Radio KZAM has adopted a program policy of contemporary music. Mostly rock. From 6 AM until midnight, 7 days a week, we fill the air with the most popular Top 40 music of the hour, news and special short features.

The accompanying diagram illustrates the breakdown of the Radio KZAM staff.

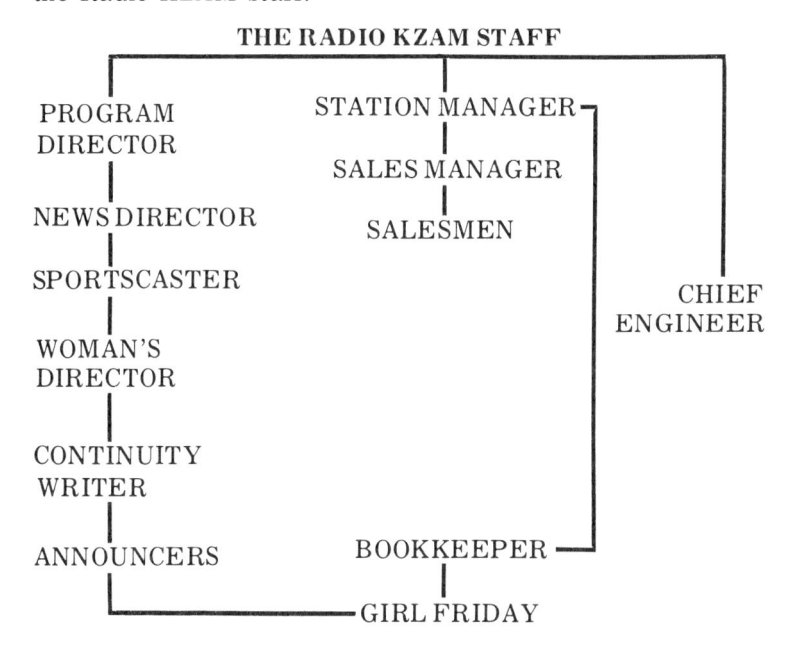

THE RADIO KZAM STAFF

```
PROGRAM DIRECTOR
  │
NEWS DIRECTOR
  │
SPORTSCASTER
  │
WOMAN'S DIRECTOR
  │
CONTINUITY WRITER
  │
ANNOUNCERS

STATION MANAGER
  │
SALES MANAGER
  │
SALESMEN

BOOKKEEPER
  │
GIRL FRIDAY

CHIEF ENGINEER
```

Station Manager: Like any other enterprise, Radio KZAM must have an executive at its head to direct and manage its affairs. Ours is George Powell, the station manager, and his duties are varied. He is definitely a man of all seasons.

Besides handling the details of routine management, Mr. Powell spends his day actively selling advertising, lunching with important advertisers, picking up news items around town and acting as a public relations and station promotion man. From time to time, George writes and voices station editorials. As manager, he is further responsible for the preparation of reports for the Federal Communications Commission.

KZAMs versatile manager started his career years ago at a 250-watter in a town of 5,000 population where he learned the tricks of his trade as an announcer, copywriter and salesman.

George Powell says, "We don't have a network at Radio KZAM. A network can't read news bulletins any more efficiently than we can. And, obviously, a network pays practically no attention to local news. With our policy of popular music, interspersed with the news, we succeed in holding our own with the local newspaper because people in the community like to hear about doings of people they know. The proof of the pudding is that we sell quite a few of our daily features to banks, dairies, bakeries and car dealers. **We produce local programming for local people.**"

Chief Engineer: Unlike other businesses, the operation of a radio station requires the services of a highly trained expert, the chief engineer. The KZAM chief, John Hatch, is responsible for its technical functioning, and he also prepares all the necessary FCC technical reports and purchases all equipment. Chief Hatch is a skilled technician who acquired his vast knowledge over a number of years of study, trial and error and "troubleshooting."

In his words, "It's a plain and simple fact that a radio station cannot be operated without an FCC licensed engineer. "In the early days, such a fellow was usually an amateur operator turned pro. I was originally a ham operator myself. But today an engineer has to be the product of a technical school who is specially trained to meet face-to-face all the complicated details that fall to the lot of a broadcast engineer. He's got to know everything that turns on, shuts off and burns out around the place."

Program Director: Larry Wayne, Radio KZAM's PD, is a combo man, holding a first-class ticket. Clearly he keeps busy with a variety of duties and responsibilities.

To start with, Larry plans the entire broadcast day. He sees to it that live and recorded programs are effectively produced, that all news elements are handled efficiently, that disc jockeys are hired, fired and scheduled properly and that the programming is produced in good taste, according to social standards and FCC regulations. Larry often writes commercial copy, thinks up station promotions and contests and works as a disc jockey himself on regular shifts. When a regular DJ fails to show up for work, he fills in as a "relief man."

It is also the PD's responsibility to keep the music library stocked with up-to-the-minute hit songs that are being played according to the station format, and he is in constant communication with record distributors as to forthcoming popular record releases and of possible changes in music trends. He reads the music sections of Billboard, Variety and Cash Box to keep on top of the music action. Larry works closely with the station manager, the chief engineer and the head of sales.

Years of experience back him up. Like the station manager, he broke into the industry as a "green" announcer at a coffee pot station in the grass roots. Larry Wayne enjoys telling how on a junior college radio station he got the "feel" for radio, and what he did about it.

"After being on the air, radio seemed an easy way to go," he explains. "I didn't want to work in a foundry all my life. From the job I held part-time, I knew it wasn't for me. So I worked hard at the school station on diction and delivery.

"One day, there it was, more than I had dreamed of, a want-ad in the newspaper: 'Wanted, Radio announcer for early morning shift' at a small station in a town ten miles away. I broke the speed limits getting there. I got the job—and also a traffic ticket."

News Director: Rick O'Neil, the newsman, acquired a smattering of experience while in high school reporting for the school paper. Later, during his tour of duty in the Army, compliments of Selective Service, he was stationed for a year at an Armed Forces Radio Service unit in Germany as a reporter-newscaster. Now, in Zamville, in commercial broadcasting Rick gathers local news, condenses and rewrites national and regional news stories that clatter in on the AP and UPI wire, and he reports it all on the air in brief 2-minute newscasts that are scheduled throughout the day. Rick works a split-shift—two 3-hour periods, 6 to 9 AM and from noon till 3 PM, Monday through Saturday.

"One of the real satisfactions of a job like this," he will tell you, "is being in the middle of everything. One day, it's an attempted break out of the county jail; the next day a Governor's press conference; or a big fire on the outskirts of town. You get on a first-name basis with the county, town and state officials."

Sportscaster: Peak radio audiences are captured by the presentation of actual live sporting events, which Radio KZAM broadcasts from time to time. But loyal audiences also follow the daily sportscasts, which are scheduled primarily in the late afternoon. Bud Barry, our man at the sports mike, uses a technique like that of the news director, combining local sports briefs with national sports news, presented twice each hour from 4 to 6 PM.

Earlier in the broadcast day, Bud voices the regular news in the absence of the news director. Bud's split shift is 9 AM until noon, and 3 PM until 6 PM Monday through Saturday.

"In school I was more serious about radio than anything," he says. "Football and basketball opened the door for me at KZAM. After I got my diploma, I convinced the station manager how valuable I'd be on the air doing the sports stuff, since I knew the subject well and all the local sports people. He gave me a chance."

Women's Director: Just two years ago, Loretta Cole, 22, was a secretary at the local Chamber of Commerce office.

Listening to radio one Sunday, she told her father that her secret hope had been to go into broadcasting.

"If you want to do it," her dad told her, "then go out and get that kind of a job."

Loretta did. She got together a resume, wrote a few sample scripts and applied for a position at Radio KZAM, using her Chamber of Commerce experience as a springboard. The station manager immediately recognized that Loretta had the kind of enthusiasm it takes to make it in broadcasting, and he hired her on a trial basis to write commercials and public service announcements.

Loretta quickly caught on to the needs of day-to-day radio operation and pretty soon she was promoted to women's director. As such, she devotes about half of her work day writing two shows, which she voices on the air. The first is a 2½-minute woman's feature, "Lady's Day" on which she discusses style trends, recipes and topics of special interest to the female audience. The second program is a one-minute feature, "Baby Bulletins," sponsored by a dairy, on which she reports news of recent births in the county served by KZAM.

The other half of Loretta's day is absorbed mainly by preparing public service announcements. She collects this information about club activities, city-wide campaigns and group projects by serving as a liaison between Radio KZAM and community organizations. She represents the station in the leading women's clubs by her active membership and attendance.

"Although I like women's news and fashion," says Loretta, "I plan someday soon to become a full-fledged reporter of hard news like Pauline Frederick of NBC. Give me a tear-gas, rock-throwing crowd anytime in place of a recipe." "**I think the broadcasting industry is at least recognizing that women can handle most assignments as well as men.**"

Continuity Writer: You know what a disc jockey does, for you hear him doing it on the air. But what about the continuity or copywriter? Ted Adams, sitting at his typewriter at KZAM, works like most continuity writers everywhere. As he expresses it, "I have to write commercials and other material by what I call **intuitive imagination.** After the salesman brings me the information about an advertiser and what he wishes to sell, I attempt to put this message in flowing rhetoric so that the DJs can read it smoothly and convincingly to reach busy people in homes, in traffic or listening to transistors at the beach. In writing each piece of copy, I visualize a person listening at home, and I have to keep the DJ's speech habits in mind while writing as well as the length of the copy."

Ted, who joined Radio KZAM after toiling for a few years as a display-ad writer at the Zamville Daily News, doubles as a disc jockey for one hour each weekday afternoon. He doesn't have a great voice; it's sort of high and hesitating but it packs enough sincerity to sell merchandise to the KZAM listening audience. And, after all, that's the kind of voice we always look for.

Announcers: Three staff disc jockeys work a 6-day week. The "morning man," Ed Thompson, jockeys a shift from 6 AM until noon, Sunday through Friday. On weekday mornings he takes a break between 9 and 10 while the program director spins the discs. "With our contemporary format, I have only a couple of minutes between musical numbers, spots and time signals, to speak to the audience," he points out. "So I try to make every second count by dropping in some sort of grabber.

"I check the newspapers and magazines for short human-interest items or funny one-liners. For example, once in awhile I toss in what I call 'Advice Bits,' things like, 'Men, if you want to get along in this world, always remember the very first lesson in the art of self defense—wear eyeglasses,' or 'Girls, men will like you lots better if you ask for very little—especially when you're buying a swim suit.' Radio audiences dig quickie material that makes them laugh, especially early in the morning to get them in good humor. But I think the real secret is to pitch it in as fast as possible, and then get back to the music."

Disc jockey Thompson originally got his career airborne by "just hanging around" one of the stations in his hometown where a buddy was employed. He began studying broadcast techniques and learning all about the equipment. In the meantime, he acquired a radiotelephone license. One day when the regular DJ failed to show up (the poor fellow was suffering from a severe hangover) they put Ed on the air in his place. Ed has been a professional DJ ever since.

The "afternoon man," Bill Morgan, commences his shift at noon and disc jockeys until 6 PM Tuesday through Sunday. From 3 to 4 weekdays, he enjoys a breather while the continuity writer fills in for him.

Bill is a product of a broadcast correspondence school and learned the basic announcer skills through home study. A help wanted ad in a trade magazine was the lead that got him his first job at an FM station. Within a few months, Bill was sending out letters of application and audition tapes to larger stations. In this manner he soon landed a somewhat better job.

Nine months later, Bill mailed a resume and taped air check to Radio KZAM. At that time the station was actively looking for a sharp DJ and Bill was invited to come around for a personal interview. He made an excellent impression and was put on the payroll two weeks later.

Bill Morgan applies a philosophy to his on-the-air work that is shared by many of today's most popular jocks. "When I first blew into town and saw all the people, I thought of how each person here has a favorite radio station, and somehow I've got to make these people listen to my station for no other reason than that they want to.

"My formula for running a program is to plan ahead to try to please everybody. I know that's said to be impossible, but it doesn't cost anything to try. I run a tight board, attempting to balance the music and the messages, and I do my best to inject personal involvement into the show.

"It hasn't taken me long to become a part of the community, both on the air and off. I make lots of personal appearances at record stores, supermarkets and teenage dances on my own time to build up goodwill. The way I figure it, in Zamville they know who Bill Morgan is, and that means that they know about KZAM, too."

The "night man," Zack Robinson, works a regular board shift that begins at 6 PM and ends with the midnight sign off from Monday through Saturday. Of all the jobs at Radio KZAM his is the most back-breaking, for he operates all alone for a 6-hour stretch each evening, cueing records, keeping logs, reading commercials, answering the telephone, preparing news, sports and weather during musical selections and generally keeping as busy as the proverbial one-armed paper-hanger. Zack does everything, in fact, that has to be done to keep KZAM operating smoothly. He never enjoys a quiet, relaxed moment while on duty.

The following is a sample of just ten minutes in the life of Zack Robinson, night DJ:

8:00-00: Identify the station, play a singing jingle and give the correct time.

8:00-30: Punch up the taped news introduction.

8:00-45: Announce that the news is presented by a local car dealer.

8:01-00: Deliver the news stories in headline form (which have been ripped off the teletype at 7:55 and combined with several local headlines).

8:03-00: Read a commercial for the sponsor.

8:04-00: Punch up a taped intro to sports.

8:04-15: Read the sports scores.

8:04-45: Punch up a weather introduction.

8:04-50: Read the weather forecast.

8:05-00: Punch up the taped news close.

8:05-05: Identify the station and give the correct time.

8:05-15: Spin a record. (During the next two minutes answer the telephone; make notations in the station logs; change a turntable not in use from 33 1/3 to 45 RPM; put the next record on the turntable and cue it; change carts in the tape player; rummage through the commercial copy and prepare for the next spot.)

8:07-15: Turn on the mike; stop the record turntable, while identifying the last record as to title and artist.

8:08-00: Give the station call letters and the correct time.

8:08-15: Read a station promotional announcement plugging a forthcoming record hop.

8:08-30: Spin a public service disc on the turntable soliciting funds for the Salvation Army.

8:09-00: Read a commercial for a supermarket.

8:09-30: Start the turntable spinning the next record, keeping the music volume low, and introduce it by announcing the name of the performer, the title of the tune and say something interesting about it.

8:10-00: Turn off the microphone, bring the music up to full volume.

Begin again approximately the same procedure as outlined above.

 Up until a few months ago Zack was a recruiter for the U.S. Army, stationed in Zackwille. In carrying out his service assignments, he wrote, produced and voiced announcements

RADIO KZAM ON-AIR PERSONNEL SCHEDULE

	Sunday	Monday	Tuesday	Wednesday	Thursday	Friday	Saturday
6 AM - Noon	Ed Thompson	Ed Thompson	Ed Thompson	Ed Thompson	Ed Thompson	Ed Thompson	Larry Wayne
Noon - 6 PM	Bill Morgan	Larry Wayne	Bill Morgan	Bill Morgan	Bill Morgan	Bill Morgan	Bill Morgan
		(9 AM - 10 AM Larry Wayne's regular show)					
		(3 PM - 4 PM Ted Adams, continuity writer's regular show)					
6 PM - Midnight	Bert Weed	Zack Robinson	Zack Robinson	Zack Robinson	Zack Robinson	Zack Robinson	Zack Robinson

Rick O'Neil, the news director, works 6 AM - 9 AM and noon till 3 PM Monday through Saturday.

Bud Barry, the sports director, works 9 AM - noon and 3 PM till 6 PM Monday through Saturday.

Bert Weed, the relief man, fills in for personnel who are on vacation or on sick leave in addition to his Sunday evening shift.

at Radio KZAM regularly where he became friendly with the staff and caught the broadcasting "bug." Just before his enlistment was terminated, Zack applied for a jock job at the station and got it.

The ex-recruiter has this to say about breaking into radio: "There's one remark that lots of would-be disc jockeys make that turns me off. They say, 'It's not what you know, but who you know that gets you the breaks.' What they're saying is that the budding DJ in question has a good talent that is brushed aside because people with less talent use personal contacts to get the jobs.

"Sure, who you know is mighty important, but I don't think that anyone of no talent or inferior ability ever succeeded in radio on the basis of personal connections. When you arrive at that golden moment when you are sitting before a mike, copy in hand and a lump in your throat, it's what comes out of your mouth that's going to make you or break you."

The relief man, Bert Weed, the owner of a successful radio parts store in Zamville and a qualified broadcast engineer, acts as KZAM's relief announcer and technical consultant. Bert DJs the Sunday evening shift from 6 to midnight, the regular night man's day off, and he often substitutes for announcers who are ill or on vacation. At other times he replaces the chief engineer when the latter is on holiday or sick leave.

Bert is a local man in his 40s whom we sought specifically for this fill-in position. We required a combo man of many skills and approached him with a job offer after learning that in past years he had once owned and operated his own broadcasting station.

"Make no mistake about it," advises Bert Weed, "the primary reason I was hired was because of my license. Announcing experience was secondary, in this case. **A first class license is the key to my career in radio. A license not only gets you a job, it allows you to be very choosy as to which job you want.**

THE SALES DEPARTMENT

The sales function of a commercial station is of the greatest importance. Everything depends on those dollars coming in from the sale of time. In very small operations everybody on the staff, from the receptionist to the station owner and his wife, sells time. The income of such a station is usually not a staggering amount, but neither is its cost of operation, and it is possible for a station in the low-power field to enjoy comfortable profits.

Small stations, in the main, attempt to serve only their own communities, competing with the local newspaper as an advertising medium for the local merchants. And these merchants have found that hometown radio is very effective in selling their products and services.

Radio is a supermarket of the air for the housewife; the best deal on a new or used car or recapped tires for the motorist; the latest x-rated movie at the Bijou for the thrill hungry; or the best place in town for a hamburger and a shake to the active teenager.

The seller of radio time, the salesman, or "account executive," as he is most often titled, handles a highly specialized commodity—an intangible. In order to sell advertising, he should know his market as well as the face he shaves every morning and the listening habits of his radio audience. He must be able to convince a sponsor that the businessman's customers are out in the bushes of radioland, and show the sponsor how these customers can best be influenced to part with a dollar. Once he has sold the advertiser, the salesman must keep him sold, and at the end of the contract persuade him to renew it.

You might say that a local salesman begins his most serious work after he has peddled the merchant a "flight" of spots or a program. This important follow-up consists of making sure that the advertiser receives on the air as much, or more, than he has been promised, and that his name and the name of his product are pronounced correctly by the DJs. Furthermore, the salesman must keep in frequent contact with his client to maintain goodwill and to demonstrate his sincere interest in the success of the advertiser's venture into radio. In short, the radio salesman has to be a jack-of-all-trades, familiar with many phases of the broadcasting business.

Most "time peddlers" get their earliest training by selling space for newspapers or through general selling of everything from almanacs to antique zithers. Sometimes they step right out of school or out of the Armed Forces and begin as junior salesmen at broadcasting stations. Others start as announcers or engineers who eventually gravitate to the sales department because the income is generally higher.

As far as Radio KZAM is concerned, our sales department is made up of a sales manager and two sales people. With the exception of accounts that are serviced directly by the station manager, these three people bring in all of our revenue. Each sales person is assigned his own "account list" of current advertisers and prospective buyers of time. These are exclusively his to call on. Through this simple system, conflict is avoided and the salesmen do not stumble into each other calling at the same merchant's place of business.

The sales staff meets once a week, usually on Monday morning, to discuss any new station policies and salable program ideas that may appeal to specific advertisers. At this meeting reports are given on prospects and on the current activities of established clients. The clear-cut theme of each sales meeting is HOW TO MAKE A BUCK!

Sales Manager: Chuck Edwards supervises the activity of Radio KZAM's sales people and he sells commercial time himself. As sales manager he reports successes (and failures) directly to the station manager.

Chuck used to be a popular disc jockey in a city of 150,000 where he sold spot announcements for his own daily show. In a short time he discovered to his surprise that he liked the selling end of broadcasting best, and he gave up air work to transfer into the sales department. A so-called "natural-born salesman," he soon built an excellent reputation.

Learning that Radio KZAM received its construction permit, Chuck applied for the sales manager position. Following his comprehensive and neatly typed letter of application, several long distance telephone conversations and a personal interview with the owners, Chuck was engaged to head up sales.

"My background as an announcer proves invaluable in selling," Chuck says. "For one thing, I am well aware of the problems facing a jock on the control board, and I keep the limitations of radio production in mind when I'm pitching an account. I never expect the man on the board to do anything that I can't sit down and do myself."

Up until last year, Bob Little was selling office furniture for Harvey's Office Supply Company. Now he is selling time for Radio KZAM. Bob made the profitable transition from bookcases to broadcasting on that eventful day when he endeavored to interest the station manager in some desks and chairs. The manager was so impressed with Bob's sincere approach that he offered him a job on the spot.

'Radio broadcasting always intrigued me, even when I was a kid," Bob admits, "Now that I'm in the industry, I hope to go on to become a full-time newscaster. I think that's my real bag." I'm working towards such a goal by coming in on Sunday evenings and doing a few live newscasts."

Marjorie Miller, an energetic gal in her middle 30s, is the third member of the sales team. She is especially adept at bringing in orders from retail stores which cater to the female trade. Marjorie had always dreamed of working in broadcasting as an actress, but in her earlier years that goal seemed unattainable. She found herself in a rut, working first at selling cosmetics and soaps door to door, and later as a salesgirl for a department store.

"I decided one day to take the bull by the horns and I arranged an interview at the radio station. Although many people make the serious mistake in an interview of saying they have no particular job in mind and will do anything from sweeping the floors to sewing buttons on the manager's jacket, and thus convincing the interviewer that they wouldn't be worth much of anything, I highlighted the one thing I was good at—selling. By not scattering my 'pitch' for a job, I made a cleancut impression. Everything is working out, thanks to my well-planned business interview. My income as a time saleswoman is very good, and I get to go on the air every now and then by taping spots for some of my advertising accounts."

Bookkeeper: Mrs. Karen Wharf does all the Radio KZAM accounting, except the annual audit. She takes care of the payroll, accounts payable and receivables. Mrs. Wharf also assists the Girl Friday in her duties, which will be defined later. And, from time to time, the bookkeeper is asked to participate as an actress in radio commercials.

Mrs. Wharf has a long and varied history in the financial field, having begun in an accounting pool at a public utility company and worked up to a top post. She advises that, "Beginners should learn about financial transactions in school and get a degree in administration, finance or accounting, if possible."

The Girl Friday: On the surface it would appear that Radio KZAM would come to a full stop without the varied skills of a gal like Arlene Jacks. Not only does Arlene take care of telephone answering and over-the-counter visitor business as our receptionist, but she handles traffic and secretarial tasks as well.

As the traffic girl, she types the program schedules on the logs in advance, making several carbon copies for control room and program department use. She sets up complete disc jockey ring-binder books for each air shift, while maintaining the files of all formats and commercial copy.

Traffic orders are scheduled by our Girl Friday, and she must be alert at keeping competitive advertisers separated from each other on the air. As secretary to the station and sales managers, Arlene takes dictation and types business letters. Her regular duties further include keeping the talent mail count, and packaging and mailing out all prizes that are won by listeners who participate in Radio KZAM contests.

"In qualifying for this job," says Arlene, "I first learned typing and shorthand at a business college while reading all the books on broadcasting that I could find in the public library. By doing this, I had a good idea of just what would be expected of me at a radio station."
Did we mention that Arlene Jacks often performs on the air? Well, she does. When the program director needs a girl with a honey-drip type voice for a station promo or a particular commercial, he calls on Arlene. Although she didn't plan it, KZAM's Girl Friday stepped onto the air via a side entrance.

SMALL-STATION RADIO

Admittedly, many of the staff people at a station in the low power field have to put in long, grinding 6-day work weeks to keep the station producing 18 hours each day, week in and week out. To be sure, this is hard work and contrasts greatly with the multimillion dollar, big-city AM down the road that can divide the work load among more than a hundred people. Yet still farther down the line, at the lowest rung of the broadcasting ladder, are found the really small facilities that send out programming from makeshift buildings staffed by as few as five persons. Why don't we tune in such a 250-watt daytimer, and check out how five people keep it going?

1. The general manager, who is in many instances the station owner, apportions his time chiefly to the selling of local advertising and servicing clients already on the air. He supplements this responsibility by writing much of the commercial copy, by gathering and editing local news, handling station promotion and public relations and supervising all the programming. And that's not all. As a qualified combo man, he disc jockeys a regular weekend shift.

2. The chief engineer, often times the second in command, divides his time between a regular daily board shift, from noon until sundown, and he takes care of the equipment and FCC reports during the remainder of the day.

3. A combo man, the chief announcer, signs the station on at sunrise and carries through on the control board until noon. Then in the afternoon he calls on local merchants to sell them advertising.

4. The Girl Friday, in such an operation, does a number of jobs. She is secretary-receptionist, bookkeeper, public service director and traffic girl. She sometimes voices commercial announcements.

5. A salesman-announcer spends most of his day contacting local retailers selling shows and spots and writing

Product demonstration on television. Host Del Gore interviews an attractive guest.

commercial copy. But he also participates as a combo man on weekends and he substitutes for staff members who are away from work for one reason or another.

In winter months, daytimers, broadcasting on a schedule of sunrise to sunset, sign on as late as 8:45 AM and sign off as early as 4:15 PM and this requires fewer on-air hours by the limited staff. But let's suppose that it is now summer with the longest broadcast days. The daytimer's on-air staff schedule could appear as shown in the accompanying schedule.

	Sunday	Monday	Tuesday	Wednesday	Thursday	Friday	Saturday
6:00 AM SIGN ON TO NOON	Combo Man	Combo Man	Combo Man	Combo Man	Combo Man	Combo Man	General Manager
NOON TO 6:00 PM SIGN OFF	Salesman-Announcer	Chief Eng.	Chief Eng.	Chief Eng.	Chief Eng.	Chief Eng.	Chief Eng.

DAYTIMER AIR SCHEDULE

At little stations with low budgets, you can see how jobs may consolidated to achieve a successful end result. Despite the distinction between the 50,000-watters with their elaborate studios and impressive staffs and the little coffee pots with a mere handful of people in a quonset hut, the style for broadcasting is basic.

And automation is another means that has come along to cut the operating expenses in radio. One lone announcer, using two-part automated units for (1) production and (2) playback can program the entire station. In less than two hours, depending on the load of commercial announcements he has to record, it is possible for him to program the entire broadcast day.

THE TV FIELD

Opportunities in the television field are most attractive, especially for beginners. You don't need a famous name to stand in front of the cameras and pitch electric mops or tomato slicers. If you can motivate viewers to buy the products you are hawking, you will be in demand by advertisers.

TV is the answer to a performer's prayer. It's show business with all the trimmings of glamor and excitement...the thrilling urgency of radio, movies and the theatre all packaged together into one hell of a bundle.

TV offers a fabulous future to any qualified go-getter who means to make it his business. Opportunities in TV are so excellent for the newcomer because talent is used up at a rate unequaled by movies or the stage. Professional movie actors hope for one movie a year, but a television performer can appear on the screen many times each day, providing his talent and training are equal to it.

Lloyd Bridges, once he dried off from his longtime frogman role in SEA HUNT, has been seen in the flesh and

Modern, clean facilities greet the KNTV visitor.

heard as a "voice over" in dozens of television commercials. Henry Fonda sells cameras, while veteran actor David Wayne speaks for McDonald's hamburgers and other national advertisers. Edward G. Robinson has stirred instant coffee coast to coast. David Janssen has appeared for headache pills. Louis Jourdan, the romantic Frenchman, has urged us to send flowers by wire. And so it goes. Dancing star Ann Miller selling both soup and panty hose; Mickey Rooney, Don Ameche, Rosemary Clooney and Rudy Valle pitching record albums.

The money is great—that's why! Top-name actors and actresses realize there are big bucks available in TV that no longer come so easily from the theatre screen. Exceptional off-camera voices rack up fabulous earnings. A good example is Allen Swift, a New York announcer, whom you have probably never seen. He reportedly commands more than $500,000 a year voicing commercials for various advertisers.

But let's talk about the opportunities at your friendly, neighborhood television station because that's where you have to start your career rolling. The big cash comes later. For starters, you should know that there are two types of TV transmission—VHF and UHF.

VHF (Very High Frequency) stations telecast on Channels 2 through 13, with Channel 1 reserved for experimental purposes. UHF (Ultra High Frequency) stations telecast on a separate band from Channels 14 through 55.

In commercial telecasting, VHF stations are universally more affluent than the "Us," with rare exceptions, just as AM radio stations are usually more commercially successful than FMs. In most instances, a UHF station runs into obstacles related to technical problems of reception and the U faces difficulty in obtaining network affiliation and, most importantly, the support of advertisers.

It is lots easier for a viewer to select Channel 4, for example, than to fiddle around trying to bring in Channel 46. This again is the name of the game—total audience and BUYING POWER. Sponsors buy advertising strictly on the basis of the number of people watching, and, stated simply, **more people** watch the Vs than they do the Us.

A TYPICAL TV STATION

I am most happy that for past years I have been in the employ of a very prosperous VHF station in California. It is KNTV, Channel 11, in San Jose, an affiliate of the ABC-TV network. KNTV-11 is a powerful V that serves over 1,000,000

In today's television station the control room is its central nerve center. Millions of dollars are invested in sophisticated electronic equipment to produce entertainment and public service.

people in one of the fastest populating areas in the United States.

By using KNTV as an example of a typical television station, we will make clear, I am sure, the basic operation of commercial TV stations everywhere. Channel 11 is not a mammoth New York, Chicago or Los Angeles type of facility. Yet neither is it akin to Channel 36, the "Sight and Sound of Buffalo Breath, North Dakota." KNTV is what I would call in-between, and the size of station where a newcomer stands the best chance of breaking into the video picture.

Seventy happy souls staff KNTV, and, in writing this book, I carried about my tape recorder and interviewed all of them. Each case history, as it turns out, is somewhat different from the other. Excellent "side door" methods of stepping into the broadcasting industry are revealed in the following pages, and should be of assistance to you in starting your own career.

ZOOMING IN

Before going any farther, it is a good idea for us to take a look at the physical makeup of KNTV, put in the simplest of terms so you won't have to plow through complicated technical phrases.

First, there is the transmitter and tower, the "stick," located atop Loma Prieta, a mountain peak, at an elevation of 4079 feet. With a power of 123.0 KW we telecast a rainbow of vivid moving colors into 12 counties, including the ultra-rich San Francisco Bay Area 50 miles away. The transmitter is miles from our studios, and pictures are carried to the mountain by microwave.

Unlike many TV stations that have gone about converting barns, warehouses and other large structures to shape make-do studios and offices, KNTV has its own specially

In the TV control room an engineer sets up a video tape reel for an on-the-air performance.

89

The technical director is the show director's right-hand man. The TD manipulates the controls at the video switching console to achieve desired special effects.

constructed, 2-story building with adjacent spacious parking and loading zones in downtown San Jose. This building was designed with television needs specifically in mind. Sometimes we call it "Gilliland's Island" in honor of Allen T. Gilliland, the station owner.

The studios and production facilities are located on the ground floor; the business offices upstairs. There is a good reason for this particular arrangement. Automobiles, campers, horses and other large, hard-to-handle exhibits make frequent appearances on TV. Thus, easy access to the street level is important. Nobody would want to take a circus elephant up an elevator!

In our main studio, 60 by 40 feet, live programming is presented, and shows and commercials are video taped. Everything is produced in full color. To keep out undesirable noises, the studio is sound-proofed by sound absorption materials on the walls and ceilings.

Since there is a necessity for a great deal of physical movement of sets and props, all such staging materials are constructed so they are completely mobile. Most sets, when not in use, are warehoused in a large area directly behind the studio.

In passing, let me mention that a visit to the studio in operation usually impresses someone who has never been in a TV station as complete bedlam. But honestly, there is genuine method in what appears to be outright madness in the constant movements of cameras, changing of scenery, wild waving of arms, cries of the floor director, technicians climbing ladders to adjust lights—a profusion of ground-out cigarette butts and mounds of empty coffee containers and soft drink cans.

The KNTV control room adjoins the studio, separated from it by a large glass window, and it is elevated above the studio floor to permit the technical crew to have a direct view of the performance. In this nerve center are combined some of the factors of a radio station control room, the shooting stage of a movie lot and the opening night of a Broadway show. Every television performance is a first night, and in the control room the program or commercial that the viewing audience is to see is directed, edited and produced. Final decisions are made in this busy room that determine what appears on the home screen.

A normal control room operation requires four specialists—the director, technical director (TD), audio engineer and the master control technician. These people not only put the show together, they control picture quality, switch from camera to camera, create special effects, regulate

91

This is KNTV's master control. The countdown equipment, multi-colored dials and pushbuttons on the operations panel provide the final step before programming goes flashing to the transmitter.

sound, put a finished product together on video tape for instantaneous replay in color and perform all sorts of "electronic miracles."

The crew sits in front of viewing screens, called monitors, which are actually direct scenes from the studio cameras, upcoming slides and film projectors. To the average visitor it probably appears like space-shot headquarters at NASA, but to the technical people involved it reveals just what is happening—or about to happen. The staff in the control room, the cameramen and the floor director in the studio all wear headsets which are interconnected to maintain constant communication.

The director, the "quarterback" of the production, sits at his "perch," which is an operations desk that permits him to concentrate on all facets of the production. He calls all the shots, telling cameramen which scenes to pick up, and cues the technical director and the floor manager. He literally puts the show together as the cameras shift from long shots to close-ups to cover all the action. Decisions are made in a flash with split-second precision, and everyone has to be quick on the trigger.

Adjacent to the control room is the announce booth, which operates practically the same as its counterpart in radio. This little room, called the "meat locker" at KNTV, is equipped with microphone and control equipment, a viewing monitor

Director Jim Risinger is pictured on the "director's perch," calling the shots for a live telecast. In the glass panel behind him are reflected mirror images of monitors picking up camera scenes that he must constantly keep in mind in order to successfully quarter-back the operation.

In the announce booth an announcer is preparing for a voice-over TV commercial.

and headset for the announcer. From this small studio come the voice-over portions of any production, and it is also used for station identification, oral announcements for slide presentations and for comments over silent film or video tape.

A breakdown of the various technical jobs appears later. We zoom in with a "word camera" to put each of them in focus for you. We explore the film department across the hall, and go down the corridor into the art department. Then into photography next door. We have a peek at the production offices and later pan the news department to see what is there. But in the meantime, as they say at the start of a studio rehearsal, "Let's take it from the top."

Upstairs.

NICE WORK—AND YOU CAN GET IT!

It is not uncommon for announcers to move into management, production or sales positions. This is especially true at KNTV. The vice president and general manager of the station, Robert Hosfeldt, is a trim, muscular, athletic fellow in his late 30s (he jogs 2 miles a day and bicycles another 3), who made the climb from radio disc jockey to top TV executive in less than 10 years.

In his position as KNTV's chief executive, he now directs and manages all station affairs. Mr. Hosfeldt's supervision entails control of financial matters, the employment of personnel, maintenance of proper relations with the federal government, broadcast unions and community organizations and the direction of the business end of the entire activity.

"It all began when I was a voice major at San Jose State College," Bob Hosfeldt explains. "I was taking a course called musicianship. You had to listen to a piano playing a single-note melody and write it down. I wasn't very good at it and the professor called me in and said, 'Bob, if you'll change your major I'll give you a C in the course. Otherwise, you're going

TELEVISION STATION KNTV-11 PERSONNEL

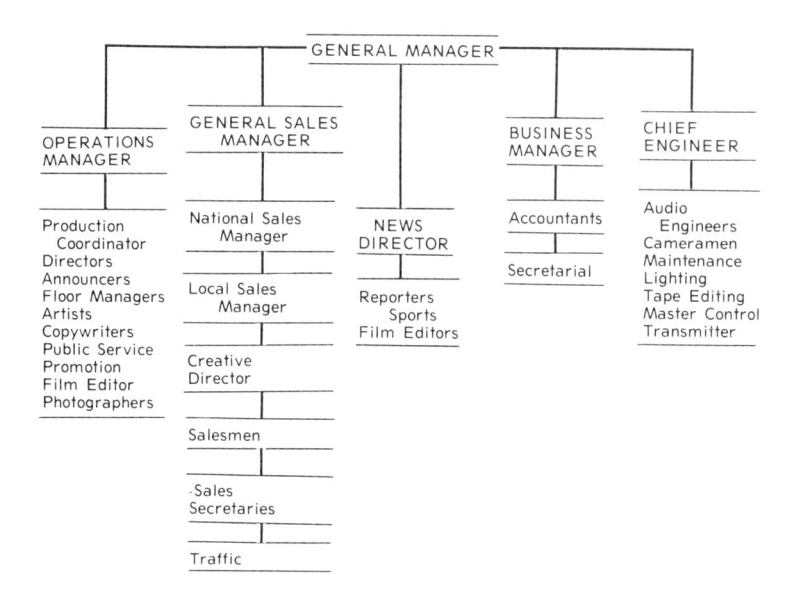

GENERAL MANAGER

OPERATIONS MANAGER	GENERAL SALES MANAGER		BUSINESS MANAGER	CHIEF ENGINEER
Production Coordinator	National Sales Manager	NEWS DIRECTOR	Accountants	Audio Engineers
Directors			Secretarial	Cameramen
Announcers	Local Sales Manager	Reporters		Maintenance
Floor Managers		Sports		Lighting
Artists		Film Editors		Tape Editing
Copywriters	Creative Director			Master Control
Public Service				Transmitter
Promotion				
Film Editor	Salesmen			
Photographers				
	Sales Secretaries			
	Traffic			

95

to get an F.' So I looked around and decided on a broadcasting major, and found out I really liked it after I got into it.

"During the summer I took a correspondence course from a school of electronics and crammed a year's course into about 90 days, took my test for a first class ticket and got it. "Then I went around to all of the radio stations in town and asked if they had any need for anyone at all.

"Finally KLOK, which was a foreign language station at that time, had an opening for a night-time combo man, and I got the only English language disc jockey show on it. I worked there full-time during my junior and senior years. When graduation time came closer, I sent out a lot of resumes to, I suppose, every TV station on the West Coast, and the only one that responded, other than a 'thank you for your interest. We have no openings, but will keep your name on file' was KNTV right in my home town. I hurried over and talked to the program director. He said he had a booth announcer job weekends. He didn't even audition me. He just took me into the announce booth and showed me how to turn the switch on, and then he left me. That was my basic training in television.

"I was going to graduate from State in June of '59, and one of my instructors who was a full-time announcer-director at KNTV got mad at the program director and told him where to go. The PD phoned me and asked if I would like to audition for the full-time job. I said I would. And so I came in for a real audition where they stood me up and had me read news copy and things off the teleprompter and ad lib. I was graduated on a Sunday and on Monday I started full-time work as an announcer-director.

"When I got started everything was live. We had a record hop which ran from 5 to 6 PM, followed by a newscast. I would direct the record hop, and then run into the restroom during the station break to put in my contact lenses, and then run out and do the three live commercials in the news. I never was able to get used to those contact lenses. My eyes would water if I kept them in over 15 minutes, so by the time I got to the last commercial, my eyes were watering so badly that I couldn't see the teleprompter.

"While I was working regularly at KNTV, I went on to get a master's degree at San Jose State. It was about that time that they made me program director, and eventually I moved up to general manager."

Keeping up his energetic studies to best equip himself for the ever-increasing management responsibilities, Bob Hosfeldt, a prolific student, has devoted much of his time learning law and accounting. For these reasons: "The

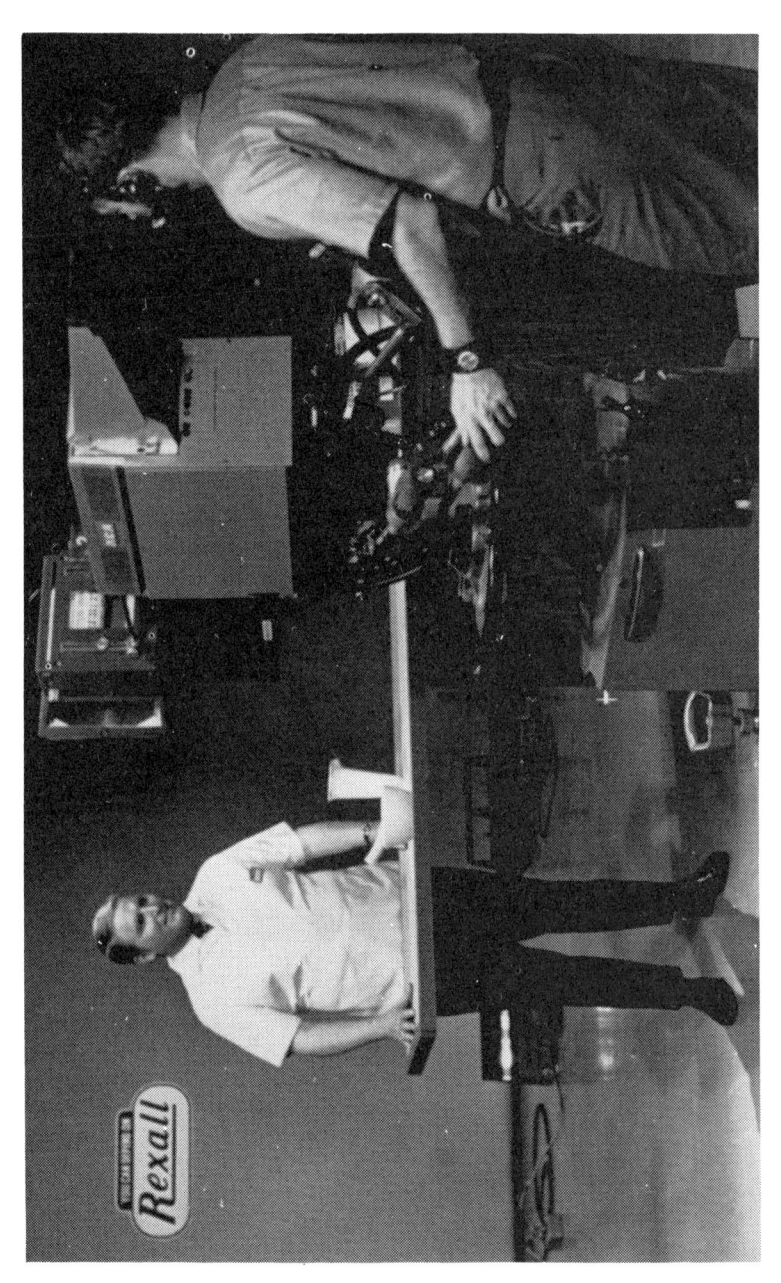

In the television studio, smocked operations manager Stew Park, a versatile performer, and an ex-radio jock, plays the role of a pharmacist in a "stand-up pitch" for Rexall Drug Stores. Guy Vaughn is the cameraman.

broadcasting business is constantly changing. You have to keep alert to that. Communications laws are changing the industry so much, a few years from now people in the business today won't recognize it. You have to be familiar with the law to keep on top of that. And a manager has to know plenty about accounting. Cost control is the most time-consuming thing I have to do. If you let that get out of hand, you are really dead."

It is the function of the operations manager, Stew Park, and his staff of production people to plan and present programming satisfactory to the management, the sponsors and the audience. As operations manager, Stew, a massive blond man, supervises the following personnel:

Director

Announcers

Floor Directors

Artists

Copy Writers

Public Service and Promotion

Film Editors

Photographers

He is responsible for the station's sound and visual signals. He sees that all copy, slides, films and video tapes are ready for daily broadcast, and he is further involved in purchasing certain production equipment and in union negotiations.

Now at age 30, Stew's case history is a good example of how a fledgling radio DJ can, in a few years, advance from spinning discs to head up an important television department.

"I developed my interest in broadcasting back in high school when I was 15 or 16," he recalls. "A friend of mine worked at Radio KSJO in San Jose, which broadcast a middle-of-the-road format. This fellow had an evening shift as a disc jockey, and he invited me to see the station. It was one of the most fascinating experiences of my life, being in a radio station for the first time. The idea that from this one room in a big city came music and talk that people all over the area were listening to and being influenced by was almost a feeling of power. I realized right there what a powerful thing broadcasting is, and I developed a real desire to become part of it.

"I continued visiting my friend, and I also worked at my announcing skills. I did this initially by just reading magazine ads and billboards out loud. When I was in a car I would make an exercise of reading all the copy on a billboard, trying to make it sound like it belonged together in a sequence like a radio commercial. I would take an ad or something in a magazine and try to string all the little blocks of copy together into a long sequence. This taught me how to think on my feet and to ad lib a little, so that before long I could pick up a piece of copy and read it so it sounded smooth.

"In this way I developed some rudimentary skills in the announcing area, and one night my friend at the radio station suggested that I read one of their newscasts. I was very scared, but, at the same time, very excited for the opportunity. I ripped the news off the AP printer and edited the copy down into headline form. And then, with my hands trembling, I sat in the center of a big studio with a microphone hanging down from a boom, a little table in front of me, and little rickety chair under me, and, for the first time, I was on the air—live. Night after night I continued doing this, just for fun and no pay, and enjoyed every moment of it. My mind was really made up to go into broadcasting.

"When I graduated from high school, I enrolled at San Jose State College in a radio-TV course. As part of the study we produced a regular series that was aired on Saturday mornings at KNTV. It was the best experience because it was very real. During my sophomore year, I realized a need for money, and I planned to get a part-time job.

"Bob Hosfeldt, who was then program director of KNTV, was going after his master's degree at State, and he was teaching a couple of speech courses there. One day I took my nerve in my hand, and, as brazen as I've probably ever been, I asked him if he would give me a job at the station.

"Much to my surprise, Bob said that they were looking for a part-time floor man, and he gave me the job. I worked as a floor man at KNTV while I continued to go to school and to practice my announcing.

"There was a teenage record hop on the air in those days that I was floor directing. The host of the show, knowing that I was interested in becoming talent, helped me along by giving me little parts in commercials that we did live on the air. I would take my headset off during the program, put on some quick costume and do the commercial with him. As soon as it was over, I put my headset back on and continued as floor manager.

"When the host was finally unable to continue with the program because he was organizing an advertising agency, they asked me to take his place. At the age of 20 this was a great opportunity. I took it like a shot and kept at it for about a year doing a lot of stand-up interviews with celebrities, and that was good for developing my ad lib skills.

"From hosting the record hop, I moved into TV directing and became what is called an 'announcer-director,' a person who directs both TV commercials and programs, and is also able to announce in a booth as a voice-over or on camera in a studio.

"After a couple of years, I was given the opportunity to become production supervisor, moving closer to the management end, directing the activities of writers, producers, artists and others and making sure that clients were correctly scheduled into production sessions. To this I have added the job of operations manager, getting more involved in policy-making decisions in terms of programming and traffic."

What advice does Stew Park, a man who hires production people, have for would-be broadcasters? He says: "In my capacity I get many applications each day from people trying to get into the business, and I find that three basic things are very important.

"**One**, when you submit an application to a TV station, by all means make sure that you have done the neatest possible work in terms of producing your application. **People are impressed by neatness.** If I read a letter from someone and there are bad spelling errors, it says a lot about that person. The sin is not in being unable to spell. The sin is not using a dictionary. This says to me that this person has not taken the time or is not smart enough to look up the word.

"It's a small thing, but you must realize that in many organizations your first contact with people of that company is not a personal contact but a **paper contact**. All they know about you is the kind of paper you use, how well you've typed the letter, how well thought out your presentation is and how you spell. The chances of getting farther than the initial paper contact depends on how well you do it. Have your application printed professionally, if you can afford it. Have a picture of yourself printed on it. Hire an artist to apply some artsy-craftsy touches. These things do make an impression.

"The second thing in applying for a TV job is to be **persistent**. I don't mean persistent to the point of becoming annoying, but being persistent in a businesslike fashion. Set up a schedule and consistently contact television stations near the geographic area that you want to get into.

"The third point is when you ask for a job be specific about the job. Never put down 'will do anything,' and try not to list four or five things if you can avoid it. You should acquire some knowledge of the business so that you know what to apply for. You're better off going after one or at most two positions for which you are qualified. Of course, you will do anything to get in, but the very fact that you ask for any job at all normally puts you in a very loose category in a man's file folder. **General Production** it is oftentimes called. Since most people who apply ask for specific jobs, a man who is hiring will look at these first. The odds are against his ever getting to the General Production folder.

"Most people get their start in large television stations as cable pullers or cue card holders or some positions that are strictly manual labor in nature. **In a local TV station, normally, the job of assistant floor man, the bottom of the totem pole in terms of production, is what you would apply for.**"

Production Coordinator. A television commercial or program, when first submitted to the production department from outside sources, may be in any form. Maybe it's a complete film with sound on it or a video tape ready for telecasting immediately. Then there are 35 mm color slides with accompanying continuity. On occasion the order appears merely as a request that something be written and filmed or video taped.

Many times what is submitted for broadcast is a mess. Not naming any names, but there exists a great number of small advertising agencies, and a handful of large ones, that are primarily print oriented. They know as much about TV production as the average individual knows about his gall bladder and spleen. Practically nothing.

That's where KNTV's production coordinator, Jan Moellering, comes in. Jan has to keep in mind the physical problems that these all present. She must consider all the angles, smooth them out quickly and get the commercial or program on the air as called for by the sales booking order.

Jan Moellering is the Girl Friday of our production department—a script girl, producer, music librarian, program assistant and lady magician of sorts all compacted into one 115-pound bundle. Add to this, the fact that Jan has one of the finest feminine voices in broadcasting, low and sincere, and we've got a gal who is in great demand. She is

101

often heard as a voice-over on many commercials telecast on the Pacific Coast for Macy's and other important advertisers.

Jan's interest in broadcasting was first sparked when she was in high school. "We had a radio arts class," she remembers, "and I always got the part of the narrator or the mother. I guess that's because I had a lower voice than most."

In college Jan majored in radio and TV production as a result of her high school activities. "Some kids who were going to school with me were doing a children's show at KNTV, and I used to go to the station with them once in awhile to help time commercials and things. In that way I got to meet the people who work here. Then a thing came along when the station needed girls to answer the telephones for a contest. They hired me, and I got my foot in the door.

"A little while later the station had a teenage dance show, and the girl they were using didn't work out. So they asked me if I'd like to take her place. Of course I did. This meant standing at the door and telling everyone to spit out their gum and making sure the boys had sports coats or school sweaters on.

"From that little job I went into answering all the mail and lining up the guests, and actually producing the show. By the time it went off the air five years later, I had found a permanent place in the station, working in traffic, typing the logs, helping in promotion and continuity and putting the copy books together. In fact, I did everything there was to be done."

Promotion & Public Service Director: The main thing Darlene Loran has to do is to promote audience interest in KNTV and its programs through press releases, ads in newspapers and TV Guide and on-the-air spots. Her job is to create a favorable image of the company. Hand in hand with this activity goes the handling and placement of public service announcements that promote worthwhile community projects, such as United Crusade, summer jobs for students, the Peace Corps, American Cancer Society and things like that.

In any broadcasting station the promotion and public service director is an important position that requires an ability to write and to generate ideas. Darlene joined the KNTV organization straight out of college where her primary interests were in drama and singing.

"I took the job in TV, figuring that to be any good in front of the cameras you'd have to be good behind them. The work had really very little to do with what I thought I was interested in at the time, but as I progressed I was given parts in commercials by our people within the station. My job gives me intellectually and financially what I need and yet still satisfies that desire to be a performer in front of the camera," Darlene concludes.

Director: Television direction isn't easy. The director must work with both sight and sound simultaneously. He sees the video on a half-dozen screens—what is going out on the air or onto video tape on one monitor—while additional screens show him what each camera is picking up, slides coming up, and so forth. In the meantime, he is listening to sounds being mixed through a loud speaker. In the midst of this complicated procedure, the director keeps following a tight script and giving terse instructions through headsets to an audio engineer, a technical director, the floor manager and his cameramen.

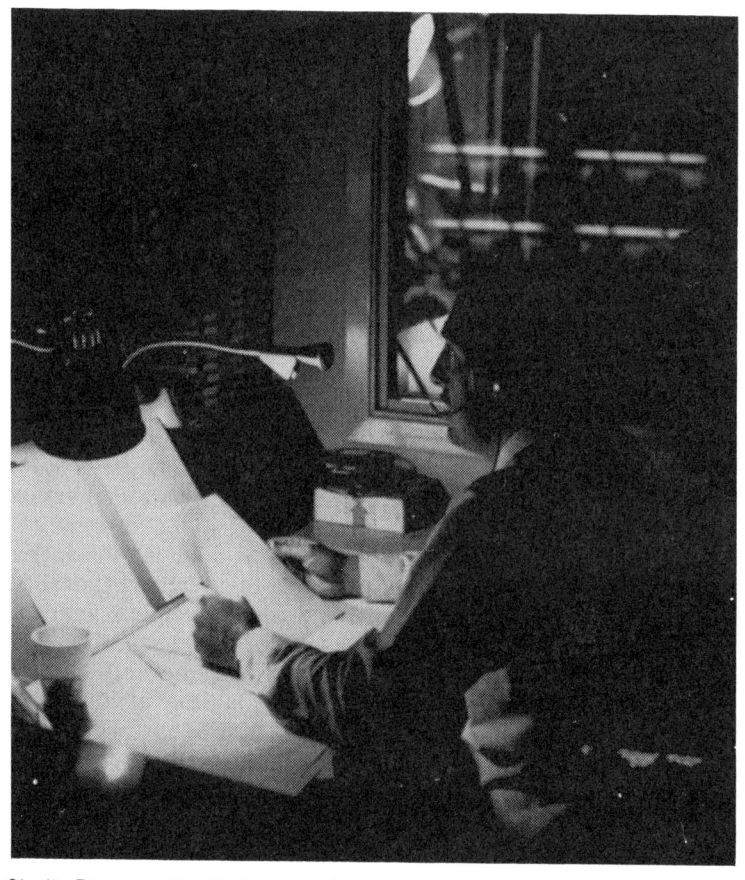

Studio Director Jim Risinger calls the shots on a live TV show. Originally Jim broke into the industry while a student working as a janitor in an Oregon station.

Despite all varieties of distractions, the director has to keep the entire production under his firm control every precious second. This calls for complete know-how of the TV medium, for speed, precision, resourcefulness, and a "sense of the show biz." A director has to know dramatic values and timing and how to best get a finished product that will "please everybody." He must know instantly which shots to use and which to eliminate. And he must do all of this in the least possible time. For, as in all avenues of broadcasting, time is his relentless competitor.

A textbook definition of a director could be put down as follows: "A person responsible for the studio production of programs or commercials and technical action. A knowledge of the theatre and of film and video tape and camera techniques are necessary. A knowledge of acting, lighting, technical equipment, film and tape editing are also essential."

So much for the textbook interpretation. A TV director is even more than that. Unlike the motion picture or stage director who can repeatedly re-shoot and re-stage during rehearsal, the TV quarterback on a live telecast has no second chance. Once the action starts, the home audience will see either a flowing performance or one that's full of foul-ups.

Jim Risinger, KNTV's senior director, faces such challenges ten times each week as the director of both the important News-11 6 PM and 11 PM half-hour newscasts, produced live and in full color weekday evenings. Jim is cool. He is a relaxed, easy-going, good-natured fellow. No matter how hectic things become just before air time, you will find him calm, cool, collected and smiling, and you can count on him for a top professional job. Had Jim Risinger chosen another career, I can see him engaged in brain surgery or paratrooping behind enemy lines.

He is strictly a graduate of the school of television experience who started at the bottom and took it from there. As Jim tells it, "Between my freshman and sophomore years at the University of Oregon, I was looking for a summer job. Things were tight that summer and about the only thing I could find was a part-time job as a janitor at the local television station, KVAL, Channel 13, in Eugene, Oregon. I was strictly emptying wastebaskets from 5 until 10. But the most important thing was, I was inside the station and could see what was going on.

"I decided this was a pretty good racket. Guys sat around and drank a lot of coffee. Since I showed interest in production, a few months later they had me working part-time in the studio. I began by running the teleprompter and helping them

Director Jim Risinger discusses with studio guests an upcoming public service announcement (PSA) for the Heart Fund. Behind the camera is Ken Willmar, a pioneer broadcaster.

move sets in and out, very menial tasks. After about six months of this, I was made a cameraman. From there I went inside the control room as a director."

Clay Pamphilon, KNTV's daytime director, broke into the trade quite differently. "I was out of school and looking for work with nothing special in mind. Since I lived next door to Bud Foster, a well-known sportscaster for KTVU, Channel 2, in Oakland, I went to his house and asked him if the station had any job openings. Bud said he'd ask. Sure enough, two or three days later he told me that Channel 2 needed someone in production and suggested I go in for an interview. The next day I got the job as a stage manager.

"After being there for a while moving props around, I started going in nights on my own time, learning to direct from the guys on duty. Then Bud Foster, my good neighbor, came into the picture again. He was involved in a new San Francisco UHF station, Channel 38, and he offered me a job as a full-time director, an opportunity I jumped at. Unfortunately, that station folded in a few months. But, by that time I was an experienced director with a few good credits to list in my resume."

Floor Director: Every production in progress must have an executive on the studio floor who receives direct communication from the director through his headset and keeps things moving. His very important function is to make sure that scenery, props and performing people and animals are all in the right positions. He cues the performers and controls the electrically operated teleprompters which are mounted on the cameras and produce script in large type that advances according to the speed at which a performer reads his lines. In short, the floor man is the personal contact with the subject matter in the studio, and he can make (or break) a production. He's got to interpret a given scene in the studio so it will look best in the viewing room.

KNTV's busiest floor director is a cheerful fellow in his 20s named Frank Casanova, whose style of hair and whiskers give him the general appearance of a young Mark Twain. His ultimate ambition is to produce national network or syndicated programs, and he is showing promise towards that goal.

Frank first got interested in show business in Tucson, Arizona when he was 13. He'd go to KTKT, the top radio "rocker" in that desert town, sit for hours and watch the DJs run the board. "Then I built my own little wireless transmitter in a closet and started my own tiny radio station," he says, "I broadcast for a radius of about 50 yards, give or take a few inches."

Before long Frank debuted into commercial radio as a guest disc jockey on a Saturday afternoon show. He next acquired that important radio telephone license and began his career on Tucson's KOPO as a jock for $1.25 per hour.

In the meantime, Frank got actively interested in movie making, studying camera techniques in his spare time and shooting everything that moved in the desert community from pony tethers to tumbling tumbleweeds. It was his experience in the broadcast medium, combined with his knowledge of camera techniques and lighting values, plus an eagerness to learn as much about the visual arts as possible, that eventually brought him into KNTV.

Announcer: The most common form of television announcer is the "voice over" fellow whom you hear and never see. In this era of specialists, he is the one who says, "This is KNTV, Channel 11, San Jose. Stay tuned for the news."

In the trade, we call such an announcer a "sprocket jockey." Like the disc jockey of radio, he is the unseen host who identifies the station, reads public service and promotional announcements over slides, introduces and signs off movies and delivers commercial "tags."

On many television stations, the sprocket jockeys come on live during the day and evening hours. On other stations, as at KNTV, they are pre-recorded. Normally, one part-time announcer voices up to 18 hours of "breaks" each day at KNTV through use of an automated system that is widely used in the television industry. The sprocket jockey sits down in the announce booth, described earlier, with a station log for the following day, a stopwatch and a loose-leaf binder prepared by the continuity department that contains all of the copy that is to be read on the air.

Working with an audio engineer, the announcer reads the continuity as called for by the station log and his accompanying scripts. As he completes each announcement, he pushes a button which puts a 290-cycle tone on the tape, which like a dog whistle is at a range too high to be heard by the human ear, but not too high for a machine. He then reads the next announcement, pushes the tone button, and carries on in this way.

Afterwards, this tape is inserted in an automatic playback unit. Once the playback machinery is set into motion, it automatically plays the taped announcements, stopping the tape when it hears the 290-cycle tones. These voice tracks coincide exactly with the video portions of the program that originate from another automated source. On the air the two come together. For 18 hours of programming, the sprocket jockey works perhaps an hour.

107

Various announcers are called upon to perform this function at our station. Most times the duty falls to the operations manager as an extra-curricular duty. Occasionally, one of the floor directors, who has an announcing background, tapes the "books." Once in a while, we bring in a disc jockey from a local radio station on an hourly basis. A knowledge of microphone technique and the ability to read easily on sight are definite requirements for a sprocket jockey.

Enter now the **studio announcer**—man or woman—and at KNTV there is a continuing parade of these stand-up pitch people in front of the cameras. They appear regularly to sell tracts of land, home appliances, draperies, correspondence-school courses, just about every product or service that you can find in the Yellow Pages.

Unlike the booth announcer who can do his work unseen with his collar unbuttoned and tie loosened and a three-day growth of beard on his face (and who's to know the difference?), the studio announcer must present himself well groomed, natural and graceful on the air. For good performance sake, he must avoid irritating facial and gesture mannerisms that will distract the viewer. No knuckle cracking, pulling of the ear lobes, or scratching of the derriere allowed!

In the first place, the studio announcer must be someone who will represent the sponsor with sincere and honest dignity. For this reason, we select the performer whom we feel is best suited for a particular spot. To demonstrate carpet sweepers, for example, or to advertise a bra and girdle sale, it is logical to select a woman announcer to extoll the virtues of the advertisers, whereas men prove best on commercials for automobile dealers, bankers, and masculine oriented merchandise and services.

Secondly, a studio announcer must be able to read copy smoothly from teleprompters or cue cards without appearing to do so, and to adapt quickly to any changes in his continuity—additions or deletions of wordage—and still keep his pitch within the specified time limits.

In rounding off any summary of studio announcers who perform at KNTV, it must be said that it is not just one person but many; whoever best fills the bill for a specific advertiser. Ah yes, to answer your unasked question, I **would** cast a dwarf in a restaurant commercial providing he was a believable on-air salesman. At least, the little fellow would make the sponsor's sandwiches look larger.

Within the station there are talented people in jobs unrelated to air work who frequently perform. Del Gore is our best example.

Del's official job at KNTV is as that of a time salesman, for he prefers selling to performing. Yet his longtime background in the industry goes back many years to the pioneer radio days when he was a disc jockey on a 250-watt station in the boondocks. In the past, Del has hosted many important local shows and has the questionable distinction of having been the first person to make a commercial pitch (for vacuum cleaners) on San Francisco television.

Art Director Warren Lamm first creates a miniature for a proposed new set for NEWS-11. This scale model is precisely detailed so that it may be studied thoroughly for workability before major construction begins on the real thing.

Silent footage and sound-on-film is processed instantly in the KNTV-11 laboratory. In this shot technician Andy Thomas is working with film that will quickly be edited and presented within the next several hours on the News-11 live telecast.

When someone is needed for an on-air commercial spot at our station, we often call on Del to "change hats" and do it, for he possesses most all the qualities of the perfect studio announcer.

(a) Del has **charisma**—a unique personality that establishes instant rapport with the viewer.

(b) He radiates **sincerity**, and appears to believe in whatever he is selling. For instance, he can speak about a patented glue that a man puts on his legs to hold his socks up with what seems genuine enthusiasm.

(c) He has **poise** and **presence of mind**. His mannerisms are natural, and yet they are slow, deliberate and economical.

(d) He is what is known as a "**quick study**," meaning that Del can usually read a piece of copy once in rehearsal and then deliver it smoothly within the proper time limit from the very instant that the red camera light winks on.

In broadcasting Del Gore is referred to as a "real pro." Just how did Del achieve a status that makes his special talents in popular demand? He says, "You combine whatever abilities you've got with the technical tools of the trade—and hard work."

Art Director: Officially, an art director is defined as "an artist employed at a television station who is responsible for planning and creating artwork for use in broadcast advertisements, public service and station promotion announcements."

Warren Lamm, KNTV's tall, bachelor art director, whom the girls call "groovy," fills the bill, and adds to it. For one thing, he is a scenic designer who plots the physical layout of a show. As a creative artist, Warren conceives the settings for productions, and he is called in to supervise activities in the preparation of sets with all the set dressings that are needed.

Processed film is put on individual reels after it has been developed in the laboratory.

111

An artist employed by a television station is responsible for creating artwork for use in broadcast. The artist in this scene is mechanically manufacturing a camera card by using the hot press.

He is additionally skilled at lettering, sketching, cartooning and building props.

You can ask Warren to create a miniature model of Noah's Ark, for example, and he will product it in double time, complete with a matched pair of mongooses, should you request that such animals be placed aboard the little craft. My office at the station is full of interesting props that Warren and his staff have created for commercials past. There is every sort of novel thing from a genuine-appearing sultan's ring that was used in a fried chicken commercial to an oversized dollar bill the size of a throw rug that illustrated values offered by a new-car dealer.

In college Warren took a special interest in advertising design and theatrical staging, but when he started with KNTV immediately upon graduation in February of 1963, it was as an assistant floor manager and part-time artist. In those days the art department wasn't the important part of the station that it is now.

As for the talent side of the business, Warren has, over the years, appeared as an actor in dozens of commercials, usually cast as a married man, successful young banker, or continental maitre'd. "A television station always needs bodies," he says. "Just being around the place guarantees that you are going to be on the air quite a bit whether you plan to or not.

Director of Photography: "A man has to try to expand his department by using his experience and knowledge to better advantage for the station," so says R. W. (Ozzie) Abolin, our head "fotog." "Try to get away from ordinary routines and add some new dimensions to your job."

Ozzie practices what he preaches, specializing in unusual effects during the shooting of still shots and motion pictures, always in color, for use by the production, public service, promotion and sales departments. Inasmuch as 16 mm motion picture film plays a necessary part in television production, Ozzie is essential to our operation. He is pressed into service daily by producer-directors like myself, filming commercials out in the field. It is lucky for all concerned in the making of films that Ozzie is an outdoors, athletic fellow whose hobbies include mountain climbing and shooting the river rapids in a kayak, because his assignments sometimes require much physical activity.

Scripts call for Ozzie to do everything from filming action in a bustling supermarket under a battery of portable lights, heckled by mobs of kibitzers, to climbing the tower at our transmitter on the mountain top for panoramic scenes. You could see Ozzie flying in a helicopter over a huge shopping center to capture on movie film the vast size of the complex. Or, you could find him filming commercials starring commedienne Shari Lewis and her hand puppets in her Beverly Hills home. Ozzie and his cameras take us where video cameras cannot. After each assignment is completed and the film developed within the station, Ozzie then edits it to the proper lengths.

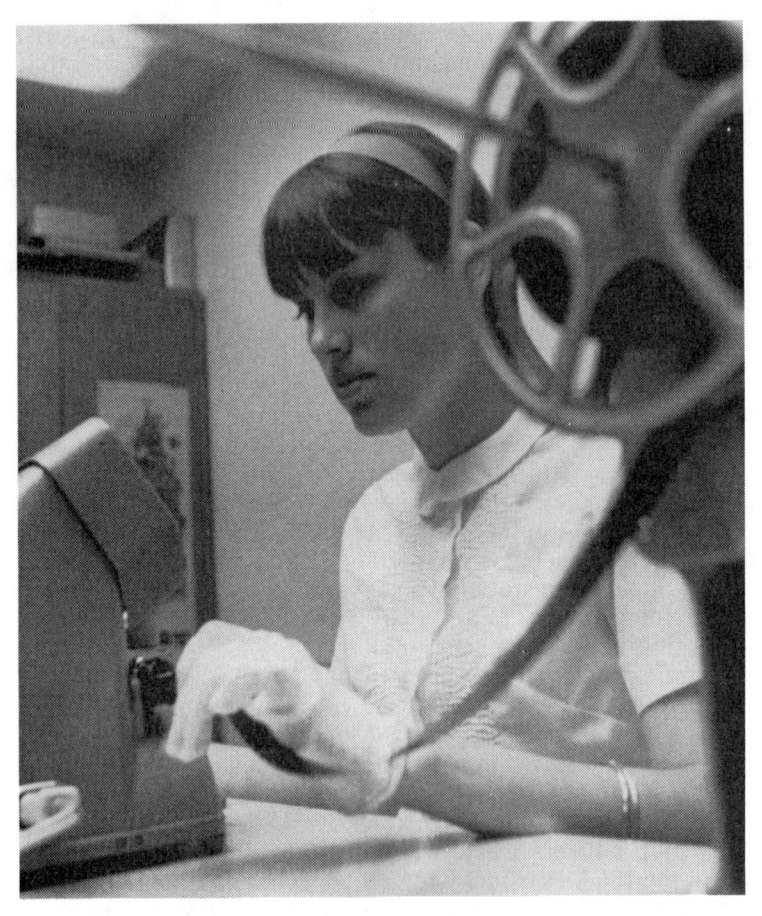

A film editor timing, cutting and splicing 16 mm film. Even the smallest television stations require at least one full-time editor, and a person seriously considering a career in film production can learn editing principles through membership in camera clubs and film societies.

In his studio, Ozzie engages in many different techniques in the creation of commercials: stop-motion photography, animation and electronic effects. He often creates humorous, off-beat effects that are not possible with studio cameras.

Ozzie entered the television profession as many photographers do. After his discharge from the Army, with the aid of the G.I. Bill, Ozzie took his professional training at Brooks, a leading school of photography in Santa Barbara. It was a 30-month course paying off in a Bachelor of Arts degree. Following this comprehensive schooling, Ozzie moved to the San Francisco Bay Area.

"I put in applications at all the TV stations. I was unsuccessful at first. But, on occasion, I would return to each of these places and ask them again if they had any openings. Finally, one did open at KPIX, the Westinghouse station." Thus, Ozzie again proved that persistence pays off!

Film Department: At some time, perhaps, while watching a movie on the "boob tube," you have been startled by a sequence in which General George Custer and his soldiers came charging over the hill, only to have the scene dissolve to a shot of comedian Jerry Lewis dashing through a crowded airlines terminal wearing only polka-dot undershorts. Actually, such foul-ups are rare, but they do happen from time to time when a TV station's film department is not on the ball and has failed to check the order of the reels.

Bud Howard and Don Ludlow, KNTV's experienced film editors, avoid such accidents by double-checking their own work and each other's under a fail-safe system. Working with viewers, splicers, footage counters, rewind reels and projectors, they carefully screen all the film received by KNTV to check sound and picture quality. Feature-length movies and syndicated shows often arrive with objectionable scratches, defective splices and even missing or reversed reels. Bud and Don must put these right.

Motion pictures must be looked at from beginning to end in advance to determine the best places to insert commercials, and often a 2-hour movie has to be expertly edited to 90 minutes. Consequently, film editors must know a picture's "story line" in order to make the appropriate cuts. In short, what film editors do is to take a film and either rearrange it or put it on the air as it is.

Looking back, both Bud Howard and Don Ludlow learned their film editing through on-the-job-training. Bud's first job in TV was in production as an assistant floor man, set designer and general flunky. Part of the work included film editing. He liked this end of the business and stayed with it. Bud is an amateur magician and a puppeteer, and these talents have brought him in front of the television cameras many times.

Don was selling hot dogs in a drive-in movie theatre in Fresno when he heard via the grapevine that KJEO, Channel 47, in that city needed a shipping clerk. He got the job. Then during the summer when the regular film editor went on vacation, Don filled in and, in that manner, started his career.

"A theatrical background helps in film editing as well as in performing on the air," Bud Howard will tell you. "I used to do theatricals in the service. Whenever USO entertainers came to the base, I always tried to work with them backstage

A view of a film editor's workbench.

and appear occasionally in front of a live audience. This is the way I learned about timing and tempo, something that's necessary in any phase of show business."

The **engineering department** is headed by Chief Engineer Lou Bell, a thoroughly professional broadcaster, whose technical-school background in Wichita, Kansas, in his youth qualified him for his first TV station position as a maintenance engineer. From that modest beginning Lou progressed steadily onward and upward in engineering. Now, as the head "nuts and bolts man" of the KNTV operation, he is in charge of all purchases, installation, operation, research, maintenance and repair of technical equipment. By all accounts, he keeps us telecasting full time 365 days a year.

Under Chief Bell are a number of skilled technicians who are directly responsible for getting the action from the cameras to the transmitter and on the air. In order that each engineer know every technical job in the place, the technicians at KNTV move about from job to job on a rotation basis, changing types of work periodically under the terms of their IBEW union contract.

Let us now describe the specialists of the control room:

An audio engineer controls the volume of the sound elements, cutting microphones in and out, mixing voice, music and sound effects in a manner similar to that of an engineer at a radio station or in a recording company.

The technical director (TD) "rides gain" on the picture elements of a program or spot. On the desk in front of him are video faders and pushbuttons that control the switching and output of the cameras. His duty is to provide, on cue, dissolves, super impositions and special effects. The TD is the director's righthand man in handling technical details.

The output of cameras, film and slide projectors and video tape machines is viewed by not only the director and the TD for its artistic and theatrical value, but by the video engineer at master control for its electronic picture quality. He is the control-room staff member whose responsibility is to deliver to the audience a perfect picture. With the countdown equipment, multi-colored dials and pushbuttons on his operations panel, he is the final authority on programming in the control room before it goes flashing on its way to the transmitter on Loma Prieta.

The cameraman on the studio floor must think fast and have a special sense of artistic picture values. When his camera is in use, the cameraman must concentrate on keeping it focused on the action, while being careful to avoid picking up any part of a set that doesn't belong in a particular scene. In general, it is the cameraman's responsibility to deliver to the director specified shots where and when they are wanted.

In studio setups, the lighting effects are under the supervision of a lighting engineer, who arranges the overhead units skillfully to improve the effect of a given scene.

Elsewhere in the engineering department are maintenance engineers installing new equipment and seeing to it that the present equipment is kept in top working condition. And lastly there is the engineer who "baby sits" the transmitter at the mountain top.

While taping interviews of the KNTV engineering staff as research for YOU'RE ON THE AIR, I turned up some interesting sidelights. I learned, for example, that some of the technicians like Gerry McKee, Al Roberts, Bill Rebello and Leroy Lovern originally made the move into broadcasting as combination disc jockeys at small stations. Other men like studio supervisor Bob Martin and engineers Tom Newman, Bill Cain, Reynold Detter, Clarence Hart and Bob Klein began building electronic gimmicks in their early school years with a

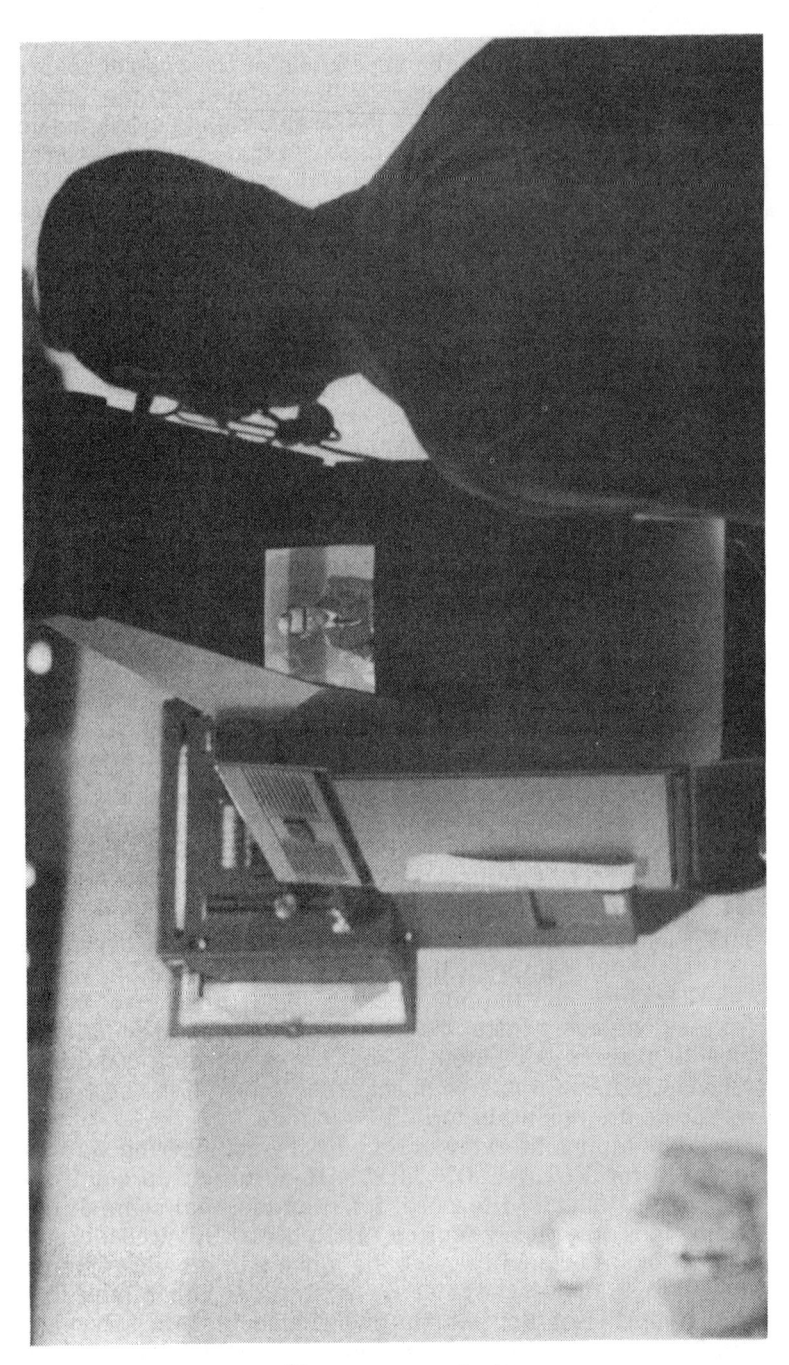

The cameraman's view.

future in broadcasting firmly in mind. Technician Ken Willman, a true TV pioneer, sold his radio repair shop and helped to construct KING-TV, the first video outlet in the Pacific Northwest.

And it was most interesting to playback one tape interview and discover that Gene Nordstrom, an engineer who, oddly enough, has never spoken a word on radio or TV commercially, has an outstanding speaking voice with deep timber in it. You can wager a dollar to a Danish breakfast roll that by the time you are reading this paragraph, people in California are hearing Mr. Nordstrom "voice over" television commercials. He made his sidedoor entrance into the talent side unexpectedly.

For the would-be broadcaster who plans to enter television via a technical avenue, Lou Bell advises, "Technical qualifications and FCC regulations require unique skills and training of broadcast engineers. But it is rewarding. TV stations employ many more engineers than radio stations and, once you get your first ticket and your foot in the door, employment is relatively steady."

News Department: KNTV's most ambitious daily undertaking is the production of NEWS-11, full-color, half-hour newscasts that go on the air each weekday evening at 6 and 11 PM. It is our thus far unchallenged claim that "NEWS-11 presents more news in a half hour than any other newscast in Northern California."

Inasmuch as a combination of many devices are needed to successfully bring a graphic picture of what is happening around the world and locally, dozens of men and women, both in front of the cameras and behind the scenes, keep on the beat relentlessly in an effort to present the news pictorially. Involved constantly are editors, reporters, writers, producers, on-air performers, directors and technicians.

Each program includes studio narration by two anchormen, a sports reporter and a woman commentator on live camera; studio interviews with newsmaking public figures and sports personalities; motion-picture film reports from newsmen in the field; and video tape inserts fed in electronically from the American Broadcasting Company. To hold viewer attention and to inject added interest, illustrations, maps and charts are also used extensively.

Fred LaCosse, KNTV's news director, is in full charge of the department, and he further appears on the air as a co-anchorman. Fred, a man exploding with nervous energy and seemingly moving in five directions simultaneously, holds a master's degree in broadcasting and his interest in the

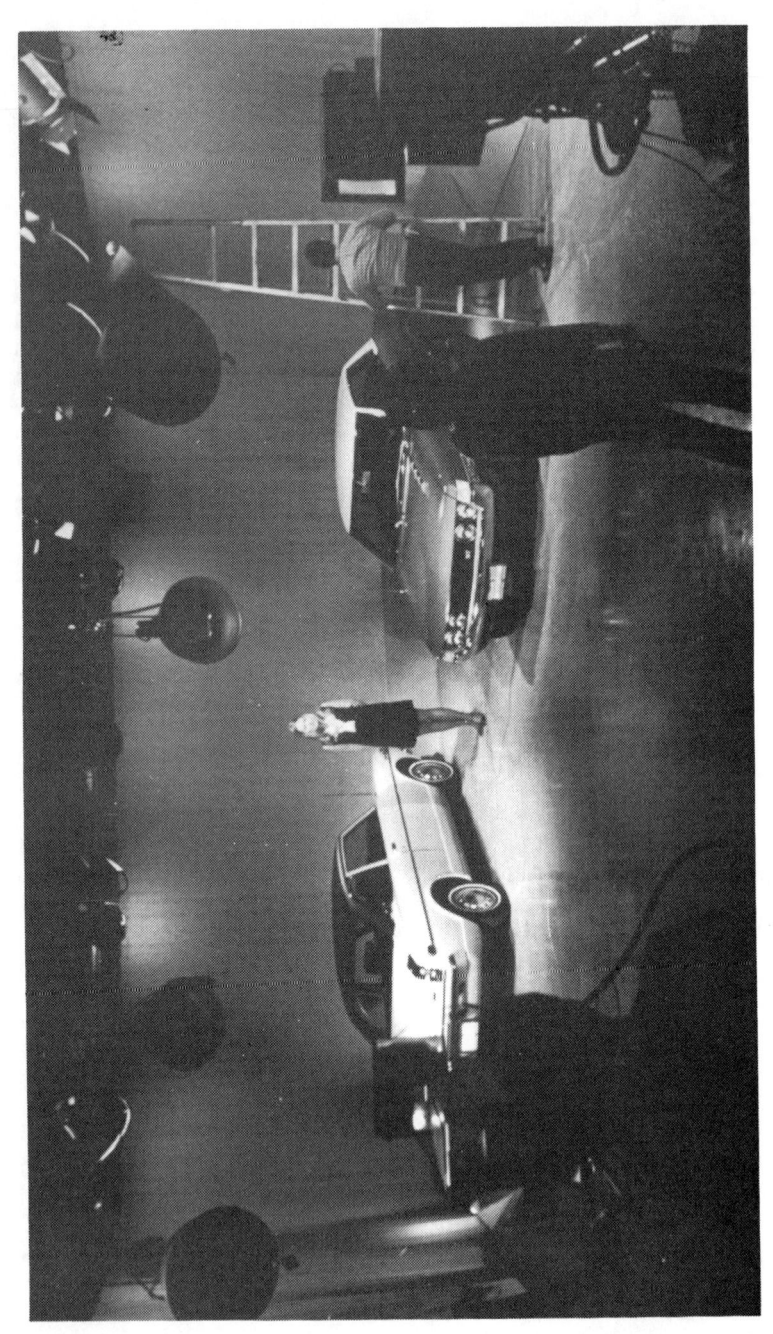

A car commercial in rehearsal. The light engineer is prepared to climb the studio ladder to fix a spotlight while other crew members stand by.

broadcast media goes back to his boyhood as a sixth grader in South Bend, Indiana, where he participated in radio spelling bees. While at Northwestern University, Fred got his first television job as a floor man for $1.00 an hour at Chicago's educational outlet, WTTW-TV. His first crack at newscasting came at WLWC-TV, Columbus, Ohio, a few years later where he voiced their 6 AM news.

As news director, it is Fred LaCosse's responsibility to manage all the activities of the news department—directing reporters in the gathering of news; writers in the writing, rewriting and condensing of stories as groundwork for broadcasting; photographers, artists, and film editors in the preparation of visual material.

"Newscasters should be straightforward," Fred says. "A newscaster should talk naturally, without the dramatic voice styles which once characterized so many broadcast announcers and commentators. Above all, a newscaster should acquire a technique in reading bulletins which create the impression that he is talking directly to the viewer as an individual without too obvious references to his script."

NEWS-11 co-anchorman Don Hayward is a versatile professional who broke into broadcasting in his teens by joining a little theatre group in Porterville, California. In one of the amateur plays with Don was a staff announcer from the local radio station, KTIP. He told Don of his plans to resign from the station and suggested that Don apply for the position on the day that severance notice was given. The plan succeeded.

Don Hayward, who became a popular disc jockey before joining the television news team at KNTV, agrees with me that the single, most important qualification necessary for any broadcasting career is learning to read aloud smoothly and naturally. "The fluent reader adapts readily to the special requirements of television," according to Don. "Once fast sight reading becomes second nature, a performer reading off a teleprompter or cue cards won't appear to be doing so. He will look natural and hold his audience more effectively. Reading from a manuscript on TV with the head down and eyes fixed on the paper, instead of looking directly at the camera, creates a disastrous effect. People will become bored and lose interest."

"A sports reporter should not over-talk; he should think of himself sitting alongside a friend in a stadium," so says John Chaffetz, KNTV's sports director. John has added new dimensions to television sports by personally involving the viewers. He conducts frequent audience-participation contests

The NEWS-11 set, as originally designed by the art director, comes to life on the air.

that give sports fans at home the chance to actually participate for prizes in competition with star athletes. Getting a hit off a big-league Giants pitcher for a color TV set is a good example.

The Chaffetz method of entering broadcast sports was definitely through a sidedoor. In Philadelphia and Washington, D.C. as a student, John used his personal charm to wheedle permission to sit in broadcast booths at sporting events and observe the style of play-by-play announcers. Later, on the West Coast, as the operator of an automobile agency, John spent many hours sitting beside well-known network sportscasters Vince Scully and Curt Gowdy watching what was going on and studying their microphone techniques. One day a sportscaster for Radio Station KEEN in San Jose realized that John was more knowledgeable on sports than most professional sportscasters, and John was given his first opportunity to voice the "color" narration of Spartan games. John Chaffetz then entered the sports-reporting field on a fulltime basis.

Viewers with a wide range of complaints call upon KNTV "Action News" commentator Sylvia Simmons to cut through the red tape and find happy solutions for their problems, something that she does efficiently as a popular public service feature within the newscast. Sylvia re-unites long-lost relatives, battles city governments for needed improvements, and goes after con-game operators with a vengeance. She will take on a federal agency with as much determination as a local butcher who reportedly weighs his thumb on the scales along with the meat. No challenge that comes to her desk seems too difficult for persistent Sylvia.

Sylvia, an attractive, dark-haired, slender woman, got her career going by voicing a weekly high school program in Cleveland some years ago. In New York, while attending Columbia University, she conducted a "Blind Date" show on WKCR, the college radio station. It was voted by students as "the most listened to program in the men's dormitories." A few years later Sylvia appeared as a weather girl on WEWS-TV, ABC for Cleveland, and for a while she worked as a show secretary for TRUTH OR CONSEQUENCES in Burbank, doubling as the on-air voice of the telephone operator who placed contest calls coast-to-coast.

Big, jovial Bob Haulman, assignment editor, is responsible for each day's apportionment of work to the news staff. He decides which stories are to be covered and who is going to cover them, and he often goes on assignments himself.

A newsroom in action. Here, news stories are gathered and put in proper order before they are presented on the air.

"Out of school I went to work driving an ambulance," Bob remembers. "On that job I came in contact with a lot of people in broadcasting who were covering stories. I got so interested in radio and television in fact that I went back to school to study it. While I was on the college radio station, the program director of KXRX phoned and asked if I'd like a part-time job three hours a day for a buck and a half an hour. The next I knew I was doing news full-time. After that I learned to fly, and I started doing traffic reports from the air."

NEWS-11 has at least half a dozen reporters in the field daily, covering local events with portable 16 mm movie cameras, shooting in color, sound-on-film.

Typical of the men who speed to the scenes of accidents and robberies, interview newsmakers, and follow up important stories is Rigo Chacon, a Mexican-American, with a strong Humphrey Bogart kind of personality. "When I was in high school back in 1962, my knowledge of English was limited," Rigo says, "so I was hesitant about approaching a public speaking course—but I did anyway. It wasn't long before I was participating at length in debates and other competition. An opportunity came about a couple of years later for me to narrate a program in Spanish every Saturday

A television newscast in rehearsal. The anchormen are discussing upcoming stories; the weather girl is preparing her weather maps; cameramen are fixing their shots; and the floor manager is relaxing.

This is the nervous moment just before a live newscast goes on the air. The Floor Manager in the foreground is awaiting his signal from the director so that he can cue the performers.

morning on KAZA, a foreign language station that catered to the 150,000 Mexican-American population of the community. From there I came to KNTV.''

Sales Department: The salaries for sales people are among the highest in broadcasting, and it must be admitted that the video venders of KNTV are not excepted from making a respectable buck. Because of the especially high rate of pay, talented performers often transfer into sales. As a matter of fact, all of our KNTV time salesmen initially got their careers going on the creative side. Del Gore, whom we discussed previously at some length, was at first an announcer and program host. Gene McGovern debuted as an actor. Bob Yochim was a commercial artist. Bill Unger once produced Miss America Pageants and helped in the development of THE LAWRENCE WELK SHOW.

These men, as sellers of television time, deal with a highly competitive and particularized commodity that must be geared to the special needs of the buyer. The fact that they must know markets, TV viewing habits, the comparative strength of program ideas and promotion techniques is the tipoff that early training at the creative end of the business is of benefit to them. In other words, if you want to be hot commercially, you have to stay cool creatively.

KNTV's head of sales is John Vera, the general sales manager. John, a tall, Kansas-born executive in his mid-forties, has a distinct style of cool. His dark, expressive eyes make you aware that busy brain cells behind them are constantly sorting and sifting a myriad of details connected with the selling of commercial spots and programs. For he must devise the best ways and means of convincing advertisers that our powerful V serves more than a million people in one of the richest markets in the United States.

The Vera career in broadcasting began at a crossroads in Phoenix in 1948 when an Arizona employment agency had two choices for him: A clerical position at the state mental hospital, or a job as a traffic clerk at KPHO. It is obvious which of the positions most interested him.

To bring sales in for us, John Vera directs three branches of the organization—national sales, local sales and traffic.

Dan McCarthy, the national sales manager, takes care of most of the coast-to-coast legwork for John by flying to major cities where he contacts national advertisers and their advertising agencies. Dan is an energetic, humorous fellow in his early 30s who has a constant twinkle in his eye.

"You need two things to qualify for my job," he grins. "A knowledge of national markets and a strong back." By the latter he refers to the load of heavy briefcases crammed with marketing data, along with a movie projector on which he shows the station's sales films, plus the luggage containing his personal effects, all of which he must pack about like a beast of burden while on the road. Dan was in athletics at Drake University (the "strong back") and later worked at the Leo Burnett Advertising Agency in Chicago for a few years assigned principally to the Campbell Soup account ("knowledge of national markets").

Jack Yearwood, the outspoken local sales manager, nicknamed "Old Silver Tongue" by station personnel, is the guiding force behind the activities of the station's four salesmen. His job is to harvest local advertising dollars by establishing and achieving sales objectives, supervising effective sales promotions and advertising campaigns and controlling commercial copy. In his early years Jack was a page, ushering visitors through NBC facilities, prior to joining a television station in Buffalo, New York, as a junior account man.

Before going farther, it is of interest to you as a would-be broadcaster for me to point out that these gentlemen make frequent appearances on the air. McCarthy, whom we think of as a "young Cary Grant," appears now and then in locally produced commercials that call for an actor to portray the typical bright young businessman. In one Father's Day promotion for Macy's, he was cast as a Dad modeling a terry-cloth robe. Yearwood, with his wide-open Irish face, and outgoing personality, plays TV parts that require diamond-in-the-rough characters. He may be spotted as an auto mechanic, a cop or a football coach. We cannot escape the fact that anyone employed around a television station is a promising candidate for on-air employment.

Traffic Department: The title of traffic supervisor at KNTV is borne by a shapely young lady named Pam Ott. In addition to keeping time charts, Pam is responsible for a dozen other odd jobs. She and her staff of girls make up the daily log, which is a guide for use by the station personnel, and they schedule all the shows, the commercials, the public service and promotional announcements.

Sales people who are interested in booking "flights" of spots request "availabilities" of the traffic department. This information is immediately available from traffic's up-to-the-minute records. As a general rule, the scheduling of spot announcements of the run-of-schedule type (ROS) is left entirely in the hands of the traffic supervisor, with the sales people merely indicating whether such announcements should be booked during daytime or evening hours.

A girl breaks into the traffic department of a TV station no differently than she might enter the fields of insurance, real estate or banking. She should have a knack for detail work and a willingness to learn on the job. Once within the station organization, however, a girl who has something special to offer talentwise can get on the air if she wishes.

In Pam Ott's case, she has three things going for her. First, she is an attractive person who looks like an average young housewife. For this reason she is often cast as such in commercials. Her other two outstanding attributes for TV work are her legs, described by dozens of pop-eyed leg watchers as "the best pair of pins on the Pacific Coast." Therefore, you are likely to see Pam Ott legs in spot announcements that advertise nylons, leg makeup or bubble bath—the sort of commercials that a producer-director (and leg watcher) like myself enjoys working on the most.

Accounting Department: Philosophically speaking, it is impossible to refer to the kindly, hard-working dollar in anything but terms of affection. The plain and capable dollar is a friend of humanity and is full of good works. Everyone loves the dollar and is hospitable to it. No other visitors are as welcome as the dollars paid to our staff when paydays roll around, and the more dollars the better. But without the accounting department, payday wouldn't happen. This important division in the company, full of lovely girls, handles all financial transactions, money collections from advertisers and disbursement for payroll, taxes and purchases.

Out of this department, as out of the secretarial pool, come TV actresses. Joanne Woodnall, an exotic beauty, is a good-looking example. She is popular both as the gal who passes out the paychecks and the lovely lass who performs Polynesian dances so beautifully on television commercials and acts out bit parts in spots for department stores, wig distributors and fashionable restaurants.

Then there is Vicki Hamer whose dark good looks and beauty-contest measurements qualify her as a natural for any video spot that requires a gal to better fill out a bikini. And we must not forget Janis Bell, whose charm and warm, easy smile make her first choice for the part of a young mother.

A girl may enter the accounting department or secretarial pool just as our own Business Manager Barbara Smith did. Barbara came to California from Canada where her first American job was as a secretary in a small TV station. Now, as a department head, she not only supervises all money matters as controller, but she serves as a personnel director and interviews job applicants. Girls who would enter broadcasting as clerks or secretaries will be encouraged by Barbara's observation: "Any city having a television station is a potential employer, and there is usually a job opening because girls get married or working wives get pregnant."

Administrative Assistant: Quite frankly, every television station should have a Woody Horn. In explaining what Forrest "Woody" Horn does, I need only to say "everything." He is KNTV's jack-of-all-trades and master of many.

Take any day. Woody is needed urgently because we are out of coffee in the upstairs kitchen. There is an insurance form to fill out and Woody has to do it. A prop parking meter is required for a commercial in progress. Only Woody knows where such a thing can be found. Where's Woody? The mimeograph machine has broken down. Find Woody! We need help carrying in a refrigerator.

I have only scratched the surface on the many facets of Woody's position as administrative assistant. But rather than write many pages explaining how he spends each busy day, let me shut off the picture on his hundreds of duties by stating simply that we depend on him to supply everything from a ceramic bullfrog to a 10-ton bulldozer.

"The more you know, the more they need you around a TV station," is Woody's philosophy, going back to his early days working with live local shows in the Midwest where his starting salary was $40 a week. Woody initially set out to be a commercial artist. But finding such jobs in short supply in Columbus, Ohio, he took a position as a floor man at WTVN-TV. There he applied his knowledge of art and stagecraft to better his position and jack up his salary.

Through the years Woody has appeared on many programs as a "walking encyclopedia" of radio and television memorabilia. Ask him, for example, how Kenny Baker, the popular tenor of the 1930s and 1940s, broke into broadcasting, and Woody will tell you that Kenny got his first paying job for $19 a week singing commercials for a Long Beach radio station. Dorothy Collins, the star of yesteryear's YOUR HIT PARADE? At age 12, Woody answers, Miss Collins won an amateur contest with a wristwatch as first prize, and then appeared regularly on UNCLE NICK'S CHILDREN'S HOUR on WJBK. Who was the Bonnie Maid hostess on the ancient Dumont Show of the early 50s? Why actress Anne Francis, of course!

It is difficult to stump Woddy with questions about the people and programs of bygone eras. Consequently, he is a natural to make guest appearances on phone-in shows where viewers hope to stump the experts. And by the way, do you need someone with a very expressive face to play a character part in a commercial? **Get Woody Horn!**

Continuity Department: The lettering on my KNTV business card introduces me as the continuity director, as does the sign on my office door. Yet I hear myself called by other names around the station—commercial producer, creative director, writer-producer, announcer-director and now and then by a descriptive expletive muttered under some disgruntled individual's breath.

Basically, my job is to produce commercials for advertisers, and often their agencies, and get these spots on the air under a deadline. This covers a handful of categories. In doing my job day to day I wear half a dozen different hats and touch base with every department in the station. By taking a hypothetical commercial that comes to my desk and following its production from start to finish, we will exemplify how a new product gets on the air.

Typically, an assignment begins when one of the station salesmen comes into my offices and says something like: "Dudley P. Peckenpaw, the pogo stick dealer, is buying a schedule of spots that begin next Tuesday. What ideas do you have for selling his popular pogos?"

At this point I must ask myself: Who the devil buys pogo sticks anyway? Just what age of audience is Dudley P. Peckenpaw trying to reach—kids, grown-ups or both? My role at this moment is that of creative director, trying to pluck from somewhere in my head a clever selling device that will appeal both to the viewing audience and the advertiser. I must decide on the most dramatic method of showing off the pogo sticks. Should we demonstrate such novelty gadgets in our studio and put the spot on video tape, or should we go out on location and shoot with movie film?

Then comes the question of who is best to demonstrate the pogos—man, woman or child? Or all three? And there is another pressing matter: How much money is the advertiser planning to spend in the production of a commercial? (Without asking, I know from long experience that this will be a much lower amount than I would hope.)

Once I have received a general cost figure from the station salesman, together with a few additional facts about the product, and it has been agreed by the two of us that a 30-second spot will sell Mr. Peckenpaw's pogos, I begin to jot down ideas on a yellow writing tablet.

Finally, I conclude that the logical approach to selling the unusual merchandise, and one that comes within the proposed production budget, is to show a dad, a mom and two children hopping happily down the sidewalk of a city street on Peckenpaw pogos. Family participation, you can call it. I write such a commercial.

Now I have assumed the duties of a copywriter, conjuring up alluring combinations of pictures and words. The central theme of this continuity has to be kept on a track, along which travels a single train of thought, one sentence after another. The sentences must move straight ahead, holding diligently to the track. After awhile, the completed Peckenpaw script comes into focus on the typewritten page.

The next step is for the salesman and I to visit the pogo stick dealer in person. During this meeting, I outline my ideas for the commercial spot and quote the cost of its production. Hopefully, Dudley P. Peckenpaw is agreeable on both copy and costs and gives us the go-ahead.

KNTV 11	**DATE** May 1 **CLIENT** PECKENPAW POGO STICKS
CONTINUITY 645 PARK AVENUE SAN JOSE, CALIF. 95110 PHONE: (408) 286-1111	**TYPE** film **AGENCY** House **LENGTH** half minute **ACC. EX.** Bob Yochim **SPOT NO.** 1

VIDEO:	AUDIO: HAPPY MUSIC BACKGROUND THROUGHOUT
	Announcer:
LONGSHOT OF SMILING FAMILY ON COLORFUL POGO STICKS HOPPING TOWARDS CAMERA	They're all together on PECKENPAW POGOS ... the family that loves to have fun!
THEN CLOSE-UPS OF HAPPY FACES OF MOM, DAD, JUNIOR & SIS. EACH MUGS FOR THE CAMERA	Everyone from six to sixty enjoys the finest hops on Peckenpaw's better brand of pogo sticks ...
FAMILY HOPPING DOWN THE SIDEWALK AWAY FROM THE CAMERA	Choose your favorite color at a department, variety or toy store near you and _jump for joy_! Just 4.95 each.
Superimpose logo: "PECKENPAW POGOS On Sale Everywhere"	PECKENPAW POGOS have come to town and fun's the limit for the entire family!

Immediately, I must swing into action and cobble together the commercial within a few days' time. From now on I become a producer in charge of all development and production of the spot. For this position, I have to have an overall general knowledge of performing, directing, technical equipment and costs.

I begin by contacting the station photographer, advising him that on the following day we will be filming a commercial for pogo sticks. Together we study the script, decide on a scenic location for the shooting and discuss the assignment in general and any problems which may arise. For best results should we film in the morning or afternoon? Is it likely that we will be harassed by a gang of motorcycle thugs in the part of town we have chosen?

Moving along, I next visit the production department where I explain to the production coordinator that I will require a video taping session in the studio next Monday. We will use so-called "dead film," I point out, edited to a half-minute, with voice and music added.

Through the art department, I next place an order for a Peckenpaw Pogo logo "key" slide that will be superimposed over part of the movie action at the time that the commercial is video taped.

Again I change hats, becoming a casting director. Within 24 hours I must have on location one man, one woman, one small boy and one little girl. These people must all be athletic enough to quickly acquire the knack of jumping about on pogo sticks and appear to enjoy it. What's more, the four of them have to look like a real family. It wouldn't do to mix up blondes, brunettes, and redheads cast together as a typical family unit. Critical viewers might get the idea that a "good neighbor" had somehow got into the genetic act.

To cast the spot, therefore, I arrange for one of our artists to play the part of the father. I then find him a suitable "wife" in accounting. And I arrange to "borrow" the two children of the general manager.

By the following morning, the preliminary arrangements are completed. The photographer, the cast of characters and I all assemble at the designated location with 16 mm movie camera, plenty of film and four pogo sticks of varying colors. In this case, we have also brought along the station salesman as a "grip" to carry some of the equipment and to perform the duties of a guard, keeping kibitzers out of our hair and thieves away from our expensive equipment. At this time I become a director, setting up the shots, rehearsing the action and eventually filming it, according to the script.

When the shooting is completed and the performers and salesman go their various ways, the photographer and I return to the station where he orders the film developed in our lab. The next day, the fotog and I screen the uncut film in a viewing room and decide which scenes to use and which to discard. With this in mind, he edits the film to its proper 30-second length and sends it on to the film editing where it is checked over thoroughly. In the meantime, from our extensive transcription library I have pulled some happy music that I feel is appropriate to the action of the spot. Briefly then, I am a musical director.

When the video tape session rolls around on Monday, I appear in the control room and inform the studio director what the spot is all about. Since I have decided to voice the Peckenpaw commercial myself, I enter the announce booth and become a "voice over" announcer, working with the audio engineer to tape a cartridge that will become the sound track of the spot. Speaking to each other through headsets, we eventually come up with an acceptable voice and music track that has a running time of just under 30 seconds.

The project is now in the hands of the studio director. He calls on the video engineer to "load" the dead film and the slide on separate projectors and has the audio engineer ready the voice cart.

The technical director cues up the high-band, color video tape machine. On a word from the studio director everything rolls—film, voice cart and video tape. The end result is a half-minute, professional appearing spot captured on a single reel of 2-inch video tape.

About this time the station salesman, who initiated the entire enterprise, arrives with the sponsor, Dudley P. Peckenpaw, in the flesh, and the video tape is screened for them in the sponsor's booth. Everyone prays that Dudley P. likes what he sees. Otherwise, I am in a heap of trouble.

Luckily Mr. Peckenpaw thinks his pogo stick spot is the greatest thing since bottled beer, and he confirms his order to telecast a flight of spots. The salesman then quickly follows through by having the advertiser sign a contract, calling for a specific number of announcements to be telecast within a designated period of time.

A booking order then goes to the traffic department where an established procedure is followed to get the spots on the air on the dates and times called for by the order. A copy of the contract is sent to the accounting department. Come the first of the month, Dudley P. Peckenpaw will be mailed a bill.

In the meantime, I have moved on to other projects. "Say, Sam, this undergarment company has begun the manufacture of sandpaper t-shirts for men with itchy backs. What ideas do you have for a commercial?"

TALENT WIRED FOR SIGHT AND SOUND

Community Antenna Television (CATV) was originated around 1950 in areas where it was virtually impossible to receive TV programs with roof-top antennas. The CATV companies put up towers and for a monthly fee connected a cable to subscribers' homes so they could receive television entertainment. During the past two decades this cable industry has mushroomed into a multi-billion dollar enterprise with more than ten million homes wired for sight and sound, and providing services not available on commercial TV.

Today there are thousands of cable systems across the country and hundreds of them are doing some kind of local programming. Hundreds more are scrambling to build studios and equip them with cameras to feed special features into millions of private homes and business offices.

This is where you can step in, because CATV has created a wide-open opportunity for performers of all varieties. Needed are announcers, actors and actresses, musicians, magicians, clowns, singers, program hosts and hostesses, technical people and men and women with salable ideas for local shows. At present the production at pioneer CATV systems is reminiscent of those first years of commercial television in the 50s. It is primitive. But remember this, out of that era came many of today's seasoned broadcasters who got in on the ground floor of a going thing. You have a similar opportunity to start your own career on a local cable system and to grow with it.

According to my son, Sam (Bud) Ewing, Jr., a West Coast program director for Viacom Systems out of San Francisco, "Local origination shows are basically neighbor-to-neighbor on the cable. Strictly local news with plenty of human interest features is the number one grabber in the CATV field.

"If there is any ham at all in your personality, there's a good chance you can make the grade on the cable. For example, how-to shows of all kinds are needed—crafts, hobbies and sports. If you are a good cook, you may be able to convince your local CATV company to put you on the cable to show off your culinary arts. Perhaps you are a well-traveled person who has hundreds of color slides depicting interesting places you have visited. An opportunity exists here for you to build a slide show which you can present on a regular basis.

"CATV has a special need for the talents of a cartoonist. You can develop a weather program or a children's show through use of this skill. Can you perform as a clown, a magician or a puppeteer? Every broadcaster knows he can capture an audience with a good program for the kids.

"The talk show format is another staple for the TV lens. Why not fill a vacuum by working up a localized version of THE MIKE DOUGLAS SHOW? Interview interesting people of your town and introduce visiting celebrities? In this area is a never-ending source of fresh, exciting programming.

"Possibilities exist for the book buff to emcee a program reviewing the latest in fiction and non-fiction. Chances are, you can talk a local book store into advertising on the cable with yourself as host. In this way everybody wins. The cable company gains a program and a sponsor, the book store increases sales—and you've got a job as a performer. The same kind of job-getting technique may be applied in creating a regular program that reviews local plays, movies and other entertainment places. "Frankly, the possibilities in CATV are fantastic, for our young industry is going places. If you have the ideas, the talent and the energy to be a pioneer, you will go right along with it."

My son warns, however, that you must not expect big money in this beginner's field. Cable-origination studios are places to start and to sharpen your skills. If performing on CATV is on your mind, so is the question: "How do I find the cable system nearest me?" Here is the plan. Look first in the classified pages of your telephone book under "Television Cable Companies," "Cable Television Systems," or "CATV." These listings sometimes vary, and you may need to query the information operator at the telephone company. A second method is to go to the library and study the CATV Source Book. This fact volume, published by Broadcasting Magazine, prints the location, ownership and management of every system in operation. In either case, you will be pleasantly amazed at the number of cable systems in your vicinity. Jot down the pertinent information, pin-pointing your prospects, and start beating the bushes.

At each cable system, ask for the man in charge of local origination, or if the receptionist gives you a blank stare, ask for the manager. CATV people are friendly and easy to approach, and you will find them eager listeners for any new program ideas, especially those that will bring in advertising revenue. Keep these three important rules in mind:

1. To fill those hours weekly, cable companies need simple, easy-to-produce programs from many sources and of all types.

2. Local interest shows draw the largest audiences on the cable.

3. Program ideas should not require a large cash outlay or too much equipment for the cable system to produce.

Typical of things to come are the plans of the San Jose Cable Company, presently in service with thousands of subscribers, and projected to be the largest CATV system in the world. 1500 miles of cable will eventually serve 100,000 or more subscribers.

"We expect to become a very significant part of the community in terms of public service and communication," says the company president Allen T. Gilliland. Initially, we will employ less experienced personnel handling educational programs. Such programming definitely lends itself to people who are breaking into the industry."

There is no denying that CATV networks are just around the corner and, in a few years they are going to need thousands of performers to appear on what hopefully by 1980 will be a national hook-up of more than 80 million television receivers. Take action now! Decide what kind of program you are qualified to handle. Put your ideas on paper. See the man at cable. This could be the open sesame for your lifetime career in a vital industry.

AD ALLEY AVENUES

In the event that there are no present openings at your local radio or TV stations or at the cable companies originating local programs in your area, your next step is to visit nearby advertising agencies and film production houses. Most ad agencies establish radio and television departments as a service for their clients and they hire talent regularly for commercial work. The film packagers contract a vast number of people themselves in the making of television commercials, sales marketing films and documentaries.

Look first at the television commercials which have taken on meaningful dimensions by creating new stars. As the result of an Alka Seltzer commercial, actor Jack Sommes, the man who appeared selling "spicy meatballs," got two movie roles quickly. Sandy Duncan was signed for both a movie and a TV series after the Pacific Coast got its first glimpse of her through a series of commercials for the United California Bank in which she played a long-suffering teller. Actress Barbara Feldon was lying on a tigerskin rug, plugging a man's product, before she did GET SMART.

On the hometown level, commercials have become a lucrative field for many and a starting point for some of the stars of tomorrow. New faces are constantly in demand to sell products. A small town car salesman who pitched his employer's rolling hardware by sitting on a ladder in TV spots became so well known to viewers that he branched out to appear for other advertisers. He makes a good living doing this on California television.

A college coed boned up on fashion phrases and convinced a Southwestern department store's advertising agency that she could serve as commentator on a TV fashion show. Wise girl that she was, she was prepared with the talent, although not the experience, to make a whopping success of the venture. She went on to find a niche in television as the hostess of a daytime movie.

In the Midwest, a salesman in a music store one day mentioned to his firm's ad agency representative that he could do a commercial playing a new electronic organ and talking about it on TV. The idea appealed to both the agency man and the sponsor and such a spot was produced. The talking organist was an immediate hit. He wound up with his own daily show on a UHF station.

So, how do you get your career motivated in this manner? The thing to do is watch television, listen to radio and read your local consumer newspapers. Decide which advertisers you might be best suited to appear for, and learn which agency in town handles the account. You can do this simply by asking the manager of the place of business you are interested in. Your next move is to visit the ad agency with a neatly prepared resume, a photo of yourself and an audition tape of your voice. Ask to see the account executive of the such-and-such account and tell him what you have in mind.

As simple as this procedure is, it is surprising how few hopefuls take this important step. Consequently, by taking it yourself you instantly better the odds in your favor. A pleasing personality backed up with the right qualifications will surely gain you interest and attention.

Quite a few organizations specialize in producing film commercials, public service announcements and programs. These independent units, sometimes called "package producers," are responsible for thousands of hours of film programming and they hire thousands of performers every year. While these companies naturally prefer experienced people, they often engage zealous beginners and are on the lookout for fresh talent.

An example is called for, and, from my own past, I can contribute an appropriate one. In the middle 1950s my partner Pat Cooney and I were employed in Hollywood by the Caples Company, a national advertising agency, as package producers of television programs in Southern California. One of our weekly shows at that time was a half-hour travel-adventure series, STRANGE LANDS AND SEVEN SEAS, telecast on KHJ-TV, Channel 9, Los Angeles and hosted alternately by movie personalities Sonny Tufts and Wayne Morris.

On this popular show was presented exciting true-life footage of intrepid explorers, race drivers, skin divers, mountain climbers, big game hunters and other adventurous souls. To maintain the high quality of the program with exceptional color action film, Pat and I were constantly watchful for anyone, professional or amateur, who could supply us with such original footage and narrate the scenes that were telecast. We spread the word through studios and cocktail bars that we were searching for people with this unique film footage for which we would pay a generous fee.

Into our offices one day came a freshly-scrubbed, smiling chap, carrying a briefcase full of materials. With him was a striking Icelandic blonde young woman with an Ingrid Berman kind of sexy voice that would steam up your eyeglasses and a small boy who had one of his top front teeth missing.

This was our introduction to Hal, Halla and little David Linker. Mr. Linker asserted how he and his family were world travelers and lecturers, and that they owned millions of feet of color movie film photographed by them in exotic faraway places with strange-sounding names. He explained that there was lots of footage of African safaris; of whaling expeditions; of Balinese dancers; world-champion Norwegian skiiers; of men who manufacture and sell illegal weapons in the Kyber Pass; of people who make sailboats that speed across the sands; and tons of other film adventures around the world. It appeared that MGM had come to us, rather than the other way around. Hal wanted to know if Pat and I were interested in using any of the films on STRANGE LANDS AND SEVEN SEAS? This was like asking a couple of squirrels if they'd like a basket of nuts!

Right at the start, Pat and I realized that the extensive Linker film library was too valuable a property to shoot into our regular program as one, two or even half a dozen episodes. The Linkers had enough material in film cans for an entire series. And that is what we made of it. We gave the show the title, WONDERS OF THE WORLD, and put together a half-hour format in which the globe-trotting Linker family would appear once each week visiting a different part of the world. The program was sold the very next day to KCOP, Channel 13, in Hollywood to fill a 7 PM Thursday night slot.

That happened more than 18 years ago and WONDERS OF THE WORLD is still a popular program, not only on Los Angeles television, but in film syndication throughout the United States on dozens of stations and cable systems. The Linkers have become national celebrities and they have had travel books published and exotic foods named after them. Yet it all began that fine day in the Hollywood office of Pat Cooney and myself. The point I am making is that you can sell an idea and yourself to an ad agency or package producer if you come around with the proper material and take the correct approach. From such a beginning, you can go far, just as Hal, Halla and David Linker have done.

FLESH PEDDLERS AND YOUR FLESH

Most cities of any size boast talent agencies, and these are presumably on the lookout for new personalities. Some of these agencies are program package producers and, consequently, have acting jobs at their disposal. However, a paradox exists, as expressed by many professionals: "You need an agent to get work, and you need work before you can get an agent." In other words, it is very difficult for a newcomer to find a good agent to represent him.

Still, it doesn't hurt to try, does it? You can ask around town and investigate the talent agencies most often recommended by performing artists until you find the energetic man or woman with whom you have a high degree of personal rapport.

Assuming you are successful in locating such a representative, the agent will want you to audition. Go prepared to do your thing! If he likes your work, the agent will no doubt offer you a contract, which specifies that he will arrange all future auditions for the people who can hire you. In return for getting you work, the agent is paid 10 percent of whatever you earn.

A word of caution: Here and there are found unscrupulous agents who will take advantage of a neophyte. They will ask that you pay them exorbitant fees. This is the time to back off, because legitimate talent agencies are licensed, and under the terms of their licenses they aren't permitted to "take you to the cleaners," financially speaking.

Thus, when you consider signing a contract with an agent, it is best to check first with the American Federation of Television and Radio Artists (AFTRA), the union of broadcast artists. The union keeps up-to-date lists of approved, franchised agents. Or, you can ask your own attorney to check the reliability of the agent before there is a chance of your being bilked.

Admittedly, an agent can be most valuable to your career, but unless you get a top-calibre one fully behind you, it is my opinion that you are better off free lancing. I once heard a stand-up TV comic quip: "I've fired my agent. Now I'm laying off **direct**."

How to Sell Yourself

PART 3

HAPPINESS IS BEING JOHNNY-ON-THE-SPOT

One of the best voices in radio and television, and one of the busiest on media commercials, was discovered as belonging to the switchboard operator at an advertising agency. She had taken the job as a means to an end after many turndowns at broadcasting stations.

A stagehand with a talent for telling jokes filled in at the last minute for a movie host who had been "taken drunk," and performed so beautifully that he permanently replaced the alcoholic emcee.

A radio engineer had just set up his remote equipment in a downtown restaurant for a luncheon interview show and was testing the line back to the station when a bank robbery and shoot-out took place across the street. The technician went on the air live with on-the-spot coverage of the action. His brilliant narration was heard by the owner of a chain of radio stations who immediately hired the engineer as a news commentator at three times his previous salary.

Every year ordinary Joes and Janes become radio and TV "names" by slipping into the industry through a side door. Therefore, your main objective is to better your odds for success by applying some simple ingenuity. **You must plan to be at the right place at the right time prepared to take advantage of the first opportunity that presents itself!**

College graduates wait in line to become network pages (ushers). Secretaries, accountants and clerks will settle for much less salary than the standard rate of pay in other fields to work in broadcasting. Highly educated young men and women, some with master's degrees, become mail boys and mimeo girls at radio and television stations, playing a waiting game. These ambitious people realize that if they want to find talent opportunities in this specialized field they must get a background of experience and be Johnny-on-the-spot when opportunity knocks. **If opportunity is seized when it comes, it will not have to be pursued when it goes.**

SEGUE TO SUCCESS

Any member of a station staff, regularly employed as an accountant, traffic girl, engineer, salesman, or copywriter, just to name a few, can move to the air by demonstrating that he or she has an aptitude for performing. From the man's standpoint, proficiency at selling, writing or accounting may be an open avenue to a spot before the microphone or camera. Understand that most local stations are small, closely-knit organizations, with staffs sometimes comprising as few as five people. Almost no stations retain a staff of full-time actors. If a voice is needed to read a stray line in a dramatized commercial or a body is required to be seen on camera, a producer or director will call on the person nearest at hand.

If you're a girl you will find that the abilities of taking shorthand, typing and running a switchboard will put you in a position where you are called on to perform. Comedienne Joan Rivers used to write material for Zsa Zsa Gabor and dream up funny stunts for the old CANDID CAMERA shows. By day she was an office worker; by night, a gratis performer in small-time clubs. But, by being close to the big-time action, she became acquainted with an NBC-TV producer and convinced him that she was funny in her own right. He gave her a chance for filling a comedy spot on the TONIGHT SHOW back in 1965. Joan's appearance with Johnny Carson at once catapulted her on to a highly successful career.

Phyllis Diller, now at the top of the show business ladder, earning millions and living in a Brentwood mansion, looks back to her first job at the now defunct Radio Station KROW in Oakland, California, as an advertising copywriter. From a hot typewriter she eventually segued to a hot microphone where modest fame on radio led to a trial engagement at the Purple Onion nightclub in San Francisco in 1955. That trial engagement lasted 89 weeks and she was on her way.

Just look at the network pages who, by being in the heart of where it happens, got the opportunity to make their names famous. The list reads like a Who's Who of broadcasting talent: Efrem Zimbalist, Jr., Gene Rayburn, Gordon MacRae, Ted Steele, Dave Garroway, Earl Wrightson and many more. Statistics show that 80 out of 100 performers in radio and television get into the business by knocking on doors at stations until they find some kind of unglamorous job in the industry. From there, they get their careers underway. The three basic steps are these:

1. A person with talent aspirations should accept any position even if it's a menial one in a broadcast operation to

get on the inside track. In other words, swallow your pride—temporarily.

2. Once employed in the industry, take great pains in learning the routine of station operation thoroughly.

3. Prepare yourself through study and practice for that grand moment when someone taps your shoulder and asks if you're interested in going on the air.

In general, luck is the idol of the idle. In your case, "luck" can be manufactured by good planning, well executed.

DISH WASHER TODAY, DISC JOCKEY TOMORROW

If someone offered me a couple of bucks to write the equivalent of the businessman's THINK sign for broadcast aspirants, I'd make it SWITCH, for this is defined by Mr. Webster as "a movable rail used for transferring cars from one track to another." In other words, entering the radio-TV field can be as simple as **switching** from one place of business to another, much like a train changing tracks.

In the field Guy Hall, a television reporter-photographer, shoots scenes of a fire.

143

As I have attempted to illustrate, all stations employ clerical and sales people, and merely by working in such a capacity at a station is a rather good warranty that the moment will come for you to perform, if you set your sights on that objective.

In broadcasting the general requirements for office work are essentially the same as for any other business. This is an entering wedge for many who later move into performance. For example, a clerk who is shuffling papers now, or adding up columns of figures for a company manufacturing 3-pronged widgets may apply for a clerical position at a radio or television plant. He's qualified.

Someone in a music store, with an extensive knowledge of syncopation from bebop to boogie, has a pretty fair chance of becoming a music librarian, cataloguing records and choosing selections to be played on the station.

Photographers experienced in taking still or motion pictures can, with some persistence, find an opening at a TV outlet. And television further needs the services of carpenters to build sets and artists to draw pictures and letter art cards.

It is possible for a salesman who greets the public easily and has some sales training in any field to enter the commercial department. After all, selling broadcast time is basically little different than selling sofas.

Switch jobs—that's the key!

You will be interested in the story of the dishwasher who became a janitor at a Texas radio station. On the surface, such a changeover doesn't appear too soul stirring, except that by working inside the station with his brooms and mops he was finally put on the air in the early morning hours by a lazy owner who didn't care to get out of bed before six and sign on the facility. The ex-dishwasher turned janitor found success at dawn as a Country-Western personality and has since moved up to the Nashville big time.

The late Bill Stern, in his day one of the nation's most famous sportscasters, took a job as an usher to get **inside** Radio City. Barbara Walters, the attractive hostess of the TODAY program, accepted a script typist job behind the scenes until she got her break in front of the cameras. Pretty Terry Lowry, a junior high school language teacher, was hired by KRON-FM in San Francisco, as a part-time translator of that station's Spanish news program. Within a few months she emerged on Channel 4 as a popular weather girl. Taken as a whole, **two things are essential to success in switching from your present job into broadcasting: One is making up your mind to do it, and the other is doing it.**

THE PERSONAL TOUCH

Horatio Alger, Jr. was the author of more than 100 novels for boys, most of which were written around the formula of a poor but worthy hero who enters life as a newsboy or bootblack, overcomes seemingly impossible obstacles and achieves the heights of success. If Mr. Alger were alive and writing today, he could certainly update his "rags to riches" formula and apply it to struggling young men and women doing unliked tasks in various fields, whose initiative puts them in the good-paying broadcast world.

First to mind comes the story of the president of a major Pacific Coast television operation. On cold mornings a quarter of a century ago, our hero was delivering milk to make a living. He did not have a particular goal in life, except he knew that carrying bottles of milk to people's doorsteps wasn't it.

Then something happened which changed the course of his whole existence and got him rolling on the route to the top of the broadcasting industry. The dairy that he toiled for sent around a memo to all of their drivers asking for volunteers to record a series of radio spots praising the quality of their milk, cream and cottage cheese from a milkman's point of view. Our hero volunteered his services, and upon being handed the commercial copy to look over during the weekend he spent many hours Saturday and Sunday rehearsing his lines. Over and over again he read them aloud. On Monday when the determined fellow faced the microphone for the first time, he performed outstandingly, and that one commercial was the key that opened his eyes to his new career. He saw a new way to go.

In the following weeks, while his spot announcement was being broadcast on half a dozen radio stations in the city, the milkman, upon completing his faithful rounds in the early afternoon, changed hurriedly into street clothing and hounded the broadcasting stations in town applying for an announcing job. His credentials for radio consisted of that one simple testimonial spot for the dairy, but it was enough. The program director of an FM station gave him a chance, and his white uniform, cap and black bow tie were hung up in his closest forever.

From that humble beginning our hero went on to become a radio and television studio announcer; then a director; and, in later years, a leading broadcast executive. In the lottery of life there are more prizes than blanks. The sure antidote for the blanks is a positive attitude, as the erstwhile milkman proved to the world.

Let's explore how his method of making it into broadcasting can apply to you in your present line of work. First and foremost, personal endorsement commercials are the very best kind. National advertisers pay thousands of dollars to housewives to endorse soaps, to astronauts appearing on behalf of gasoline products, and to athletes urging kids to eat specific breakfast foods. The testimonial spot is as old as broadcasting itself, but it is still as fresh and as popular as the minute it was first presented.

Let us imagine that you are employed as a garage mechanic and that the company you work for is a buyer of television spots. In your spare time, work up a spiel of just under a minute in which you, dressed in the overalls of a mechanic, demonstrate how automobiles are repaired at your garage. Open the hood of a car, pointing to the engine and explain how efficiently and economically you and your fellow mechanics take care of automotive problems.

After you have developed a professional appearing pitch (and your mirror will reflect the merits of your presentation) take the idea to your boss and perform your testimonial commercial for him. Tell him you will appear on television without pay and for the good of the company. Unless your employer has stones between his ears, he will accept your offer if you make a creditable pitch.

After that, you will be strictly on your own, but remember this: Almost everyone has in him the essentials of greatness if he would learn the necessary steps. It is better by far to use a shortcut through which others, like our friend the milkman, prosper and attain their objectives.

Certainly it is possible for the testimonial gimmick to quickly put you on the air. Shoe repair man, car or appliance salesman, floral arranger, insurance agent, even mortician—in any field of endeavor it is easily applicable. If you work in a clothing store and look good in your merchandise, who better to appear for the store than yourself? As a bag boy for a supermarket, you can devise a believeable commercial by rolling a shopping cart full of groceries on camera and mentioning your store's economy specials of the week.

Do you work in a fast food drive-in? Go on the air and tell of the quality ingredients that go into every burger and milkshake and point out how a delicious meal may be purchased for less than a dollar. Whatever your present occupation, you can put yourself in the commercial picture and make a mark for yourself practically over-night by applying the personal endorsement device.

Success wears many faces. An analysis of the careers of hundreds of outstanding people in radio and TV discloses that six elements are responsible for their success:

1. Knowledge that they had the ability to perform.
2. An earnest desire to succeed.
3. Perseverance.
4. Intelligent preparation and effort.
5. The ability to influence people.
6. Enthusiasm.

If the company you are working for is building better mousetraps, you're the one to tell the world about it!

SPECIAL PROMOTION

Hide not your talents; they for use were made. What's a sun dial in the shade?

Benjamin Franklin

UNACCUSTOMED AS YOU ARE...

If you are reasonably knowledgeable of any interesting subject, you may be able to convince your local station that you are qualified to conduct a daily or weekly program that the public will enjoy. Many nonprofessionals have become successful on-air performers reviewing books and movies, hosting cooking shows and conducting physical fitness programs.

Bob Wilkens, a copywriter for an advertising agency, volunteered to KTVU in the San Francisco Bay Area his own brand of low-key humor as master of ceremonies of a Saturday evening CREATURE FEATURE program. Wilkens' formula was to lampoon weak, low-budget horror pictures, like "Attack Of The Tulip People" and "Naked Came The Vampire," and to apply an "instant replay" to some scenes. He selected the most gruesome moments or significant examples of bad acting for rescreening after the film's end.

Bob further spiced the show by inviting amateur film producers to display their creations, and he interviewed a

Arts and crafts are popular with TV viewers. In this scene, host Del Gore works with guest "Crafty Molly" in showing how novelty wall plaques can be made by do-it-yourself buffs.

series of guests who claimed to be witches, werewolves and residents of other planets. The program became a matter of pride and high ratings for KTVU, thanks to the host's unusual approach and acquaintance with horror movies. It is a format for you to give some serious thought to. Perhaps such an idea could be sold to a TV station in your own area with yourself as the movietime emcee.

Another TV possibility exists in the do-it-yourself show. If you are handy at crafts, you may construct such a program in which you demonstrate on the visual medium how various things are made using simple materials. There is a good chance for you to tie-in with a local hobbystore for full or part sponsorship. Cliff Arquette, better known as "Charley Weaver" on HOLLYWOOD SQUARES and other national programs, began his career years ago by hosting a do-it-yourself feature.

The "Between Us Gals" format is always a salable commodity, and the success of such a venture will depend on your ability to dispense worthwhile tips on losing weight, new fashions, cooking and things of that nature.

The man who has a sound knowledge of sports can create an interview show, bringing to the studio important sports figures to capture a "jock strap" audience. And here is a real challenge. No one has ever put together a successful TV format "For Men Only." There are all kinds of sports programs which have masculine appeal, but nothing that covers manly arts and hobbies. If you have "something different" running around in your head, now is the time to tackle it and take it to your local television station.

TV has a special need for the cartoonist's skill, and a number of performers got their earliest breaks cartooning on kids shows, weather programs and doing chalk-talk commercials. My good friend George Lemont, one of San Francisco's most popular performers, and, in my opinion, one of the quickest wits on earth, was at first a draftsman. He made the transition from the world of machinery and structures into television on KRON-TV appearing as a cartoonist-clown on the CARTOON CIRCUIT childrens' show when the regular host was summoned into military service.

George enriched his act by the addition of exciting hand puppets whose funny dialogue and comic realism appealed equally to kids and grownups. His humorous little characters—Scat the Cat, Rhode Island Red the Rooster, and Dynamo Dudley, the Used Robot—became celebrities in their own names. You have probably watched them speaking individually for advertisers, and they frequently appear in syndicated comic strips.

The "between us gals" type of show is a popular staple for daytime television audiences. There is always a demand for interesting guests.

You may break into broadcasting as a Christmastime Santa Claus. Providing you are round in stature (for TV) with a jolly, "ho, ho, ho" kind of personality, there's an opportunity for you to capture a St. Nick role on the air in the weeks just prior to Christmas. In early November usually, radio and TV stations approach local merchants to sell Yuletide advertising packages. The Santa Claus show is perennially a winning sales tool and an audience grabber. And it is almost universally true that stations prefer someone not familiar to their audiences to play the Santa Clause role.

A letters-to-Santa program has become a popular staple on radio and is formatted in several ways, depending on station policy and equipment. Sometimes Santa, in a booming, jovial voice, merely reads letters sent in by small fry asking for Christmas goodies. Another routine used by radio programmers is to invite the kids to phone in and speak to Santa in person. The hosting of either type of show does not require very much experience in ad libbing. Children and their parents are never too critical of Santa during the holidays.

On television it is somewhat more difficult to play the part of Father Christmas, because you are in view of an audience most every second. But a simple format of showing new toys, interviewing little boys and girls in the studio, and reading a few "I want" pieces of mail from children will see you through. Stores which feature large toy departments are excellent prospects for sponsorship of such a program. If you are a woman, please do not feel that you are left out in the December cold. You can become a "good fairy," "Magic Lady," or even "Mrs. Santa Claus."

In passing, it is of interest to note that at the time George Lemont was a draftsman for General Electric, and first bitten by the "TV bug," he practiced after working hours in front of a "dead" camera that G.E. had on exhibition in its lobby. "I felt kind of goofy doing that," George admits, "but it was a good self-help way of preparing myself for the real thing."

AIR BREAKS

If you have any special talent, by all means, let your local station know about it. If you can bark like a dog, you may be in demand one day for a dog food commercial. Or for a Thanksgiving promotion if you can gobble like a turkey. Television actress Mary Jane Croft, Lucille Ball's neighbor in the HERE'S LUCY program, did especially well financially speaking for "Cleo," a basset hound in THE PEOPLE'S CHOICE starring Jackie Cooper.

Come to think of it, here is a novel approach for you that can pay off beautifully: Put on the tight green habit of Robin

Children's shows usually capture a large audience of the small fry. Here, we see Hocus Pocus, the TV clown, with a group of children on a live telecast.

Impersonating famous people may be your forte, and often this is a shortcut for getting on the air either as a guest on an interview or variety show or by mimicking famous people in radio promotional spots or commercials. Comedy star Art Carney began his climb to fame when he was 18 and named the "wittiest boy" in his senior class. His hobby was impersonating celebrities. This led Art to audition for bandleader Horace Heidt, who hired him for $50 a week to travel with the orchestra and do mimic bits. After four years as a comic on the road with Heidt, Art got a spot on a weekly CBS news program impersonating Franklin Roosevelt, Winston Churchill and Dwight D. Eisenhower. In time, he became a full-fledged radio actor on the old daytime soap operas and went on to spectacular successes in the entertainment industry.

Allen Swift was a child when he saw a movie of the late Maurice Chevalier. He began imitating the French actor's voice for relatives and friends. Today, Swift is a voice specialist for radio and television commercials, earning a reputed half-million dollars a year.

In programming television the most obvious source of entertainment consists of talented artists doing whatever it is they do best. If you are a musician, for example, your music could be the way in. Glen Campbell is a good example. From the age of 7, Glen was playing guitar with his brothers on a small time radio show in the Ozarks. At 14 he left home to join a nightclub band in Texas. For seven years he played small clubs in the Southeast. By 1962 he was a guitarist in Hollywood. Then on to the top.

Ted Mack, the year in, year out host of THE ORIGINAL AMATEUR HOUR, broke into show business as a jazz musician. Super-star Fred MacMurray originally played saxophone with popular traveling orchestras, and he used to sing through a megaphone Rudy Vallee style.

When performer Wayne Newton was 10, he and his brother Jerry auditioned at KOOL-TV in Phoenix where they became stars of a local program for years. On the whole, anyone with a splash of talent—musician, magician, singer or comic—can locate a professional place within the sphere of broadcasting. So don't be bashful.

Program directors of radio and TV find that **the unusual is actually the usual** in the day-to-day creation of entertainment. These men would not think it too surprising for you to appear at a broadcasting station reception desk costumed as a 16th Century bard and carrying a lute. If the program merchants have need for such a minstrel, they will sign you up in a jiffy.

Hood with a feather in your cap, grab your trusty guitar, mandolin or uke and go see the leading used car dealer in town. Explain to this startled gentleman that you will sing his praises on the local television station as "Robin Hood giving away good deals" from his place of business.

Ridiculous, you say? Don't knock it till you try it! Morey Amsterdam played a "flu bug" in a TV commercial. John Roberts, Sally Betts, Carl Livingston and Eddie Cohen were "raisins." Johnny Baylor was the "rear end of a horse."

An old adage expresses the point well: **Use whatever talents you possess. The woods would be silent if only those birds sang that sing the best.**

LOCKER ROOM TO CONTROL ROOM

When the discovery was made some years ago that athletes could best communicate with sports-minded audiences, revolutionary changes took place in broadcasting. Smooth, announcer-type voices gave way to Southern accents and Western drawls, much bad grammar, and a "just plain folks" style of narration. The media lovingly opened its doors to the athletic fraternity and invited them on the air to describe events involving the use of muscles. It has come to pass that nowadays most sports announcers get into radio and television through the back door by making a transition from the arena to the broadcasting booth.

An exception to the rule is the case history of noted sportscaster Howard Cosell, who claims that he "snuck up on them" by putting his early legal training together with standard techniques of play-by-play announcing to establish rapport with the fans. But Mr. Cosell did enter broadcasting via a sidedoor. He was an attorney in 1953 and received a telephone call one day asking for permission to use the name of Little League, one of his clients, for a summer replacement radio show. Better still, inquired the caller, would Howard himself emcee the program by getting kids to ask provocative questions? Howard would—and he did—and so successfully that the show ran for more than five years. Soon he was managing ABC Radio Sports. Howard Cosell's emergence into sportscasting, notwithstanding, most men at the sports mike make the leap from ballplayer to broadcasting.

For example, Curt Gowdy was an outstanding basketball player at the University of Wyoming and got his first shot at sports announcing on a small Cheyenne radio station. Football opened the door for Frank Gifford at CBS. Baseball got Sandy Koufax on NBC. Tony Kubek, who played every position for the New York Yankees, except pitcher and catcher, was asked to join NBC's baseball broadcasting team following an injury in a touch-football game. Joe Garagiola is an ex-catcher, whose emceeing of big sports affairs attracted such favorable attention that he was signed by NBC.

If you have made some kind of mark in local football, basketball, golf, track and field, skiing, bowling or tennis, there's a good chance for you to elbow your way into broadcasting. The experience you have gained from sports will equip you nicely for a future at the microphone.

As a matter of course, the shrewd station operator seeks a knowledgeable sportscaster familiar with the intricacies of scoring, the identity of various players, the meanings of the officials hand signals, and the implications of any given action on the playing field. An ordinary, run-of-the-mill staff announcer doesn't qualify to do a competent sports reporting job.

Furthermore, you can't fool an audience. Fans intuitively spot a legitimate sports buff and relate to him as opposed to someone who has little feel for athletics. Authentic sportscasters possess a mysterious quality that shines through and puts them in tune with the listener. For these reasons, the station must look to the game fields for a "jocko" personality.

If you have been a popular athlete and have the makings of a sports commentator, if you are willing to work at the trade, and if you know what is expected of you, you should be voicing sports in a short time. But, just for safety's sake, buy yourself a little "insurance."

1. Augment your present know-how of playing, drawn from personal participation, by studying the rules and procedures of each sport. You can get and study those small, inexpensive rule books published by sporting goods companies. From these you will learn correct terminology, exact dimensions of the fields and courts, pointers on playing each game, and important statistics.

2. Read every word in the sports section of your daily newspaper every day—AND READ ALOUD. Keep practicing your delivery and diction towards building up a distinctive broadcasting style.

3. Before you approach your local station, looking for a position, create a sports program of your own **on tape** to present at the interview. Such an audition program should be approximately 12 minutes in length and consist of the following components:

(a) An original program title.
(b) A format which will remain the same from show to show.

(c) Sports news, including some scores.
(d) The "inside story" on some local team or player.
(e) A brief interview with a local sports personality.
(f) Predictions of the outcome of upcoming games.
(g) Any short special features that you can think of to add zest to your presentation.

After you have impressed the manager of the station, and he gives you an opportunity to perform at the sports mike, remember these last words on the sportscasting subject: **You must become professional on the air the same way you did on the playing field—by starting from scratch.**

STAGE FLIGHT

We must not overlook the much ignored little theatre. I say ignored, not because amateur theatricals aren't widely attended and appreciated by millions of people, but because 99 out of a hundred fledgling radio and TV performers don't seem to know very much about this wide-open training ground.

Next door, down the street, or across town from you there are, I am sure, active little theatre groups. As someone has quipped, "If a farmer fills his barn with grain, he gets mice. If he leaves it empty, he gets actors." This is to say that there are innumerable amateur drama groups all over the country putting on stage plays that provide good experience for the newcomer. And your appearance in such presentations can prove a fast trip into professional broadcasting. Television producers often look to the community stage for new faces.

In my own case, while casting television spots in Northern California, I frequently call on Armand Plato, a local little theatre director, to send around talented stage actors and actresses to audition for parts in commercials. Sometimes I ask Armand for a particular type of character voice that is needed to narrate a spot. From his amateur talent stable have come harem girls for a hamburger video ad, a carnival barker personality to sell used cars, a 17-year-old actress to play a fairy princess and show with her magic wand the marvels of an enclosed shopping center, and others. As a matter of record, little theatre many times is the very quickest route into broadcasting. It's good training, and you have the chance of being seen and heard by people who can hire you.

Actress Connie Stevens, when a teenager, joined a repertory company, took a couple of years in drama and got her first TV role through a commercial she did for a bakery. Comedian George Carlin, known widely as "The Hippy Dippy Weatherman," while in the Air Force stationed in Shreveport, Louisiana, joined an amateur drama group and there met the owner of a local radio station who put him on the air as a disc jockey.

Paula Prentiss was seen by an MGM talent scout while acting in a nonprofessional stage play. She was given a screen test which led to movie roles and a television series. Howard Duff was plucked from a Seattle repertory playhouse performance and hired as a staff announcer by Radio Station KOMO. Later, he brought the original SAM SPADE to radio and progressed to starring motion picture roles and the FELONY SQUAD video series. LAUGH-IN comedienne Ruth Buzzi began inventing her characters while studying acting at the Pasadena Playhouse. She created "Gladys," for example, as her conception of the inept Agnes Gooch of "Auntie Mame."

Many pages could be filled recounting the case histories of famous performers who forged their careers from modest beginnings on the amateur stage. But let us think of how you can shoot into the limelight.

It is especially easy to get started in theatricals, for little theatre companies welcome the ambitious beginner. An amateur play producer casts you in a production because he likes you and because he thinks the audience will like you. An established name isn't important to him. The experience you gain on the stage will be invaluable. You will learn how to interpret scrips, develop timing, learn how to judge audience reaction, and how to best "sell" lines.

Now to find a starting point. If you live in an average-sized town, there are many working amateur groups which will allow you to develop poise before an audience. If your hometown has a community theatre, that's obviously the place to head first. Aside from being technically nonprofessional, these are educational and recreational, and only a small percentage of the thespians engaged there are seeking the professional spotlight. For this reason you will not have too much competition from the membership of actors and actresses trying to find a wedge in the broadcasting media. You will find that the radio and television people who do participate in amateur theatricals usually do so as a hobby. In this instance, you can cultivate a station contact that will be valuable to you in seeking station employment.

Community organizations, like the Lions or Kiwanis, often produce plays by using local talent. Then there are societies for drama in organizations like the Jaycees and the Four-H

Clubs. You can learn from your local Chamber of Commerce just who these groups are, and how you may join them.

The next thing to do is, DO IT!

After you have a little acting experience under your belt, and you feel that you are primed to make the move into broadcasting, send free tickets for your next performance to TV and radio station people and to those advertising agency executives who sign up talent. Your nonprofessional appearance on stage, thus, becomes an actual audition that can lead to a professional job.

Professionally speaking, a small role is better than a complete loaf.

PUBLIC EXPOSURE

Working as a performer anywhere, even if it's for free, is better than not working at all. It is your way of developing the power to communicate to an audience and of bringing yourself to the attention of broadcast executives. Make, rather than avoid, opportunities to speak in public. The more nervous you are about speaking, the more important it is.

Other than little theatre productions, which we have discussed in some detail, there are all sorts of other opportunities right in your own backyard, so to speak, where you get up and talk before audiences. To gain experience, you can offer your services to your local church group, a woman's club, or to any organization that presents shows of any kind. Volunteer to act as master of ceremonies for an upcoming show or dance. In assembly halls and classrooms there are often chances to introduce guest speakers or to give your own speech. And there are societies for drama, music, oratory and the like in which you can participate.

Most every town has a variety or discount store which employs men and women to announce special sale-priced items over a public address system every few minutes. This is an effective means of acquiring some good experience while getting paid for it. A high school student can spin discs at record hops, moderate quiz programs and take part in school forums.

If opportunity offers, enroll in a class in public speaking. Here you will be with a group where everyone is as interested as you are in improving his ability to speak in public.

Whatever experience you pick up, the contacts you make and your exposure to a live audience will be to your advantage when you apply for work at a commercial station. You will never regret the time and the effort it takes, once you enjoy the satisfaction from your acquired ability to capture and hold the attention of a group of people.

PUBLIC SERVICE ANNOUNCEMENT

The man who is born with a talent, which he is meant to use, finds his greatest happiness in using it.

Goethe

THE MODEL EMPLOYEE

Someone has said that a model is a girl who shapes her career around a good figure. And I would hasten to add that any city having a television station is her potential employer. Beauty and poise pay off on the tube, as witness Bess Meyerson, Julia Meade, Carol Lynly and countless other lovely females who got their first foot in the door by modeling. There always is a good solid demand for beautiful girls in TV, and the model-actress who wishes to make a living out of it can do so nicely for a reasonably long time.

The average model has believeable good looks, which are not overbearing or artificial. She has good manners, good grooming, habits, and a pleasant personality. Most importantly, the model-actress has **individuality**. In other words, she does not copy the hairdo styles and mode of dress that are glamorized by famous screen actresses and are the "in thing." **She is herself.** To be able to enter TV, a model should not be a shrinking violet or too much of a ham. Either trait will hold her back. It is essential for her to strive for a controlled middle-road personality.

Assuming that you are a talented model between the ages of 17 and 30, how should you go about putting your glamor on the television screen? First, your future as a performer begins with knowing yourself.

Does your figure measure up?

Your face?

Do you have poise and self confidence?

Is your voice quality good?

Your diction?

159

Can you read lines in a script with a smoothness of delivery and expression?

Your potential auditioners are going to judge you on the fore-going characteristics, so take inventory of your capabilities now and develop them as best you can.

Before presenting yourself for an interview at the television station, dress and groom yourself carefully. Know that you look your very best.

You must be prepared to leave with the man at the station a carefully prepared detailed resume of your past experience, education and business references, if any. You should include recent 8x10-inch glossy photographs that show you off to good advantage in street clothes, in a bikini and in an evening gown. List your physical statistics. (Hopefully, these are outstanding.)

Plan in advance the important points you will want to bring out in your interview. If you have done your homework carefully, you will know a great deal about the local shows that are presented on the station and the commercials that are produced there. With such research in mind you can, perhaps, suggest to the station executive in a calm, businesslike manner, ideas in which he can use your services in current productions. You won't need to flutter your eyelashes or show a lot of nyloned leg if the presentation of yourself is cool and controlled with selling points in it.

Be honest. Lying about anything, such as making up mythical jobs you have had, seldom works, because the people in television keep up on what's happening in the business. One thing that may help you to get a part is any **special ability** that you have. Dialects come in handy. So does singing and dancing. If you speak foreign languages, ski, sail or swim expertly, tell the man.

As to just who **the man** may be varies from station to station. Thus, when you present yourself to the receptionist, be courteous and patient at all times, remembering that her job is often a trying one because at a TV station she deals from morning until night with a "mixed bag of nuts." Explain to her that you wish to see the person who hires the talent for shows and commercials on the station.

This can be one or a combination of people—operations manager, program director, writer-producer, studio director, creative director or maybe the general manager. By calling around from department to department, the receptionist will in time find a first contact for you. He may be the wrong man to talk to, and you could be shuttled on a merry-go-round from office to office, but in the long run you will face someone who can put you to work on camera.

Believe it or not, you will find that the average interviewer will make every effort to put you at ease. His job is to learn things about you, and you will be pleasantly surprised at the interest he shows in your potential. A clever interviewer is usually a pretty good amateur psychologist who realizes that if you are relaxed you will be more informative. So, smile. Be easy to talk to. Let him evaluate your personality and weigh the facts about you. **You may not realize it, but the impression you make during an interview is as carefully noted as the outcome of any on-camera audition.**

You now ask, "What can come of such an interview?" The worst thing that can happen is, of course, **nothing**. After you leave his office the interviewer forgets all about you. Your objective, to overcome such an eventuality, is not let him forget. Phone him every few weeks, remind him of the interview, and ask if anything has come up in which you may take part. It is surprising how often this mild "bugging" pays off in getting commercial work.

The best thing that can happen during an interview is for you to get an on-air assignment immediately. You may have approached the man at just the right moment in time. In the briskly paced, ever-changing business of television, this is a frequent occurrence.

There is a third middle ground in which an interview has a delayed action; yet a profitable result. From my own experience as a writer-producer at KNTV-11, I can cite a specific example. As producer of many commercials at the station, in 1971, I was given a photo resume of Jan Gregg, a popular fashion model, whose exceptional face and figure characteristics are striking enough to light up a pinball machine.

Height: 5'8"	Size: 8-10
Hair: Blonde	Bust: 36
Eyes: Hazel	Waist: 23
Excellent hands	Hips: 35
Spokeswoman	Shoe: 8-B

In a low, liquid voice that could melt steel, the beautiful Miss Gregg told me that her ambition was to branch out from fashion modeling into television. She had much experience as a fashion show commentator, she said, and believed that this background would benefit her on the television screen. I agreed. Unfortunately, at the time she called on me, nothing was doing commercially in which Jan's talent and rare good looks could be utilized. So, into my talent file went her resume and photos to join dozens of other presentations from hopefuls.

A few weeks later I was approached by the ad agency account man representing Aamco Transmissions. He wished KNTV to produce a series of commercials for that company in which a shapely actress would appear as "Miss Tender Loving Care," a lady doctor for ailing automotive transmissions. The Aamco girl would dress in a light blue top with a red heart on it, striped red, white and blue hot pants and white boots. Most needed, said the ad man, was a girl with plenty of sex appeal whose face and figure were unfamiliar to television viewers.

Out came my voluminous talent file and, much in the style of Peter Graves on MISSION IMPOSSIBLE, I laid out various photos of models on the top of the conference-room table. Jan Gregg's picture glittered from the montage of beauties just as the brightest star shines forth in a skyful of stars. We chose her as "Miss TLC," and the commercials she appeared in have been telecast in many parts of the country.

The biggest thrill, any casting director will tell you, is to discover a new face or a new voice. Just calling in people who are already around has no real kick in it. What all producers hope for is the person who walks right into the office for the first time and bowls us right over. You could be the next model-actress to do it.

AND NOW...THE NEWS

Pick any small town in America, and you will find that most of the folks there get the lion's share of their news from radio and television, the same as the rest of us. But reporting in the heartland is a world apart from what goes on at rich networks and metropolitan stations with their large staffs and mobile equipment. Let's look.

By and large, broadcasting outlets in the smaller areas are handicapped by lack of budget, staff and equipment. They must depend on the wire services and their network affiliations, if any, to take care of most of their news. And yet, it is the **local items** that they need to liven up their newscasts. That's where "stringers" come in.

By definition, **a stringer is a correspondent, living usually in any outlying area remote from the broadcasting station, who feeds news items to editors in the central market.** A stringer may be a housewife, student, school-teacher, or, in a word, anyone. These correspondents phone or mail in news stories. Or, in the case of TV stringers, each has a sound or silent-film camera with which to cover stories pictorially, and they ship their film to the station via the fastest means, usually by commercial bus.

Opportunities for stringers are attractive, especially for beginners. If you can supply a station with the kind of audio and-or visual reports they want, they will buy your services even if you've never had a single whit of experience in the news-gathering field. On a part-time basis, to begin with, imaginative reporters and photographers can help the station make up for its deficiencies in personnel with a practical solution to the gathering of strictly local news. Stringers often make it onto a station's payroll at top salary when they prove they are reliable reporters.

What are the premises for covering the news as a stringer? To make people as aware as possible about what's going on around them in your town. You relay hard facts, chase sirens, and cover city government activities.

News may be gathered on the scene, described and-or photographed (movies or still shots) as you see it. If you arrive after an event has occurred, you can interview eye witnesses and then prepare a second-hand account. Then there are some events, such as public rallies or protest demonstrations, which are purposely staged to provide material for news stories. You can plan in advance your coverage of these. When you, as a stringer correspondent, have prepared a report of the event, you submit it immediately to the broadcast editor by putting the telephone to work or by rushing your film to the television station.

Likely sources for news coverage are city hall, law enforcement offices, and the Chamber of Commerce. On election nights, correspondents are essential for reporting the latest tabulation of votes from out-of-town precincts.

Lots of reporters score by phoning in the results of football, basketball or other games to keep a station's audience informed up-to-the-minute on the sports picture. In ski areas, a stringer can make quite a name for himself by supplying advance information of road and snow conditions for weekend skiers.

Although working as a news stringer does not require special knowledge or experience, other than a general acquaintance with good grammar (and photography, if required), there are some basic techniques which you will find helpful.

First off, you must remember that the content of the material you report is the editor's chief concern when deciding whether or not to use it. Your reports have to be newsworthy.

They have to be accurate. Keep the facts straight. The items should be kept short. That's the way to make them sparkle.

You must avoid anything repulsive or disgusting, or anything that is slanderous. Never use swear words. Never use anything offensive to any race, religion or ethnic group. Make sure the names of people and places are correctly pronounced. Be businesslike in gathering the news and businesslike in submitting stories to the broadcast editors.

To get started as a stringer, I would suggest that you go in person to the news editors of out-of-town television and radio stations that serve your area. Submit to these men your qualifications and show your eagerness to represent them, pointing out how you can supply news they can use from a town in which they do not have adequate coverage.

You can grab the attention of a news director right off the bat if you take along a carefully prepared report on tape or film of a newsworthy event that you have covered and which is still timely enough for broadcast. This can be something like an interview with the Podunk police chief in which he says that bicyclists are becoming a hazard to auto drivers and will be run into the pokey if they don't start observing the rules of the road. Regardless of what your subject is, it will illustrate to the editor that you mean business.

At scores of regional radio and TV stations around the nation, small-town stringers every day are supplying coverage of the events of their own locales, using legwork and imagination. Many of them will go on to become reporters and personalities on big-city stations.

AUDIENCE PARTICIPATION

Another effective way to earn public acclaim and come to the attention of broadcasters is by appearing as a "professional guest" on talk programs. Frequent guest shots on mini, localized versions of the David Frost, Merv Griffin or Mike Douglas shows can lead to your steady employment by the station or to a program of your own. I have seen this occur many times.

In its early days, television attempted to program entertainment from a resident company of performers. But the programmers soon realized that the public tires quickly of the same faces, and that the source of supply of interesting people has to come from outside the studio walls. As a result, we began to see on the screen a never-ending parade of guests with gimmicks—the inventor of a rocking chair with a seat belt on it; the author of "How To Avoid An Emotional Crisis

While Selecting French Pastry"; a happy male window cleaner direct from an engagement at the Y.W.C.A. Yes, as far-fetched as these.

Today there are so many talk shows on the air that producers oftentimes find themselves hard pushed to come up with someone out of the ordinary. Television, always desperately hungry for bright new faces and fresh material, will put on the air almost any interesting guest from a plastic fruit maker to a termite inspector. In that classic talk-show format with chairs in a semi-circle, you meet people, people, people morning, noon and night.

After close evaluation and from personal experience, I can make a strong case for guesting on these talk shows. By way of example, I was once assigned to produce a one-hour daily live talk program on KNTV. At the outset I decided that an interesting weekly feature would be a review of new books. Towards this end, I met with Paul Ohran, the manager of a local outlet of B. Dalton, Bookseller, and proposed to Paul that each Monday morning at 8 (the toughest time of the week to obtain guests) he come on the program and tell our audience of the latest in fiction and non-fiction and appraise these books. I convinced him of the merit of such an engagement—excellent free publicity for the bookstore and an interesting feature for us.

From the very first show, public reaction to Paul's book feature was astonishingly good, and his appearances on the program were increased to three times weekly. Paul was such a solid hit, thanks to his intimate knowledge of his subject and professional handling of the program segment, that in a short time he was asked to write a regular book column for newspapers. Next he expanded his activities to radio.

There's no reason why, as a returning guest on a local talk show, you can't make a professional breakthrough just as Paul Ohran and so many others have done. It is necessary only to have enough "smarts" on a particular subject that will excite a sizable segment of the viewing audience.

But before we explore the **how's** of your doing it, let's look at a second case history, that of Alvin "Junior" Samples, the 300-pound overall-clad fellow of HEE HAW. Junior got out of a sawmill and into the entertainment world simply by telling tall tales.

A game warden came by one day and asked Junior about a large fish-head lying on the flatbed of a truck. Junior related to the official a wild story of how the huge fish had been caught, and the impressed game warden relayed the fantastic yarn to the conservative Georgia Game and Fish Commission.

165

The next thing Junior knew a disc jockey from a Macon radio station heard about the report and interviewed the country boy for his program. Junior's story was played on the air a few times, and suddenly a record company bought rights to it and released it as "The World's Biggest Whopper." To everybody's surprise, it was a best-selling disc and Junior Samples became a celebrity. Pretty soon Junior was starring on HEE HAW and moving his wife and six children out of a $25-a-month shack into a $50,000 home with three bedrooms. From small beginnings giant careers can grow.

In your town, perhaps, or close by for sure, there is a television station that programs a talk show. Citizens appear with pet peeves. A shoe merchant displays the latest in ladies' footwear. A movie personality tells about his latest film. A policeman reports on burglaries and how to prevent them. The chef of a smart restaurant reveals the recipe for a banana surprise. Guests come and go day after day. I will put money on my prediction that three things are true of the program:

1. There are several regular guests who return periodically—an astrologer, physical fitness exponent, artsy, craftsy person, expert on dieting, or whatever.

2. One (or more) of these "professional guests" will ultimately spin off into a program of his or her own.

3. The producers of the daily show are on the lookout for a new feature to incorporate within the program.

It is my experience as a longtime producer of many shows that good guests are hard to find. During all my waking hours, I read newspapers, listen to radio and watch the TV news and special features, searching for interesting people with offbeat hobbies or occupations—a lady sky diver, rare bottle collector, belly dancer, palm reader—who can bring something novel to a program. Believe me, these persons do not come to the program of their own volition; they must be discovered, contacted and invited to appear. Surely this gives the hopeful an edge in getting started.

Take inventory of your particular assets. Do you play a musical instrument? If you do, why not organize a combo and offer to perform on the program on a regular basis for no pay?

Can you do comic bits? A comedy routine somewhat like Charley Weaver's "Letters from Mama" could secure for you a continuing slot on the show.

As an amateur ventriloquist, magician, puppeteer or cartoonist, you can combine this talent with a patter of topical humor. Jerry Van Dyke first got public exposure telling second-lieutenant and chow-line jokes in the Air Force. This experience led to his own local television show in Terre Haute, Indiana. Don Knotts performed monologues in high school and got started in show business with a ventriloquist routine. A magic show by Carsoni The Great helped prepare Johnny Carson for his first announcer job at WOW, Omaha.

I hear you saying that you have no special talents, except talking. All right then, let's work up a straight talk format for you. Astrology is always a popular subject, especially with daytime women audiences. You can guest on the program, giving a rundown of the day's horoscope for the ardent viewers. So many books have been written on the zodiac that you will have no trouble preparing material, and this is a perfect vehicle to insure your continuing appearances day after day.

Format a "How It All Began" featurette. From information gleaned from encyclopedias and a wide variety of books that can be found in any library, it's relatively easy to develop fascinating accounts of ancient customs, superstitions and habits that still influence us today. Using appropriate props, it's possible to cover everything from the origin of the alphabet to the reason for the pawn broker's symbol of three golden balls.

A pleasing daily stint is "It Happened On This Day." From your research, pouring through back issues of newspaper files in the local publishing office, you can construct a short feature that recalls what historical events occurred in years past. If you appear on the program on December 13, for example, you could relate that on that day in 1642 New Zealand was discovered; George Gershwin's "An American In Paris" was performed for the first time in 1927; and Sam Ewing, author of YOU'RE ON THE AIR, was born in 1920; and so forth. You would put the greatest stress on major events, but you should go to some length to include human interest, **localized** incidents of bygone days, as well as some humor, whenever possible.

The clincher for guesting on a local talk show is to invent a more or less original short feature that can be quickly prepared and presented in such an interesting way that it has universal appeal.

EDITORIAL

Sweat + Effort = Success

167

IT PAYS TO WIN FRIENDS
AND INFLUENCE PROGRAM DIRECTORS

Contacts in the broadcasting business are most important, so cultivate friendships in the industry. By making the acquaintance of station people, one day you may experience the break you are looking for.

Art Hannas, a prominent CBS newscaster and Ed Sullivan's program announcer for 16 years, made his first breakthrough into radio when he was a youth in Olean, New York. In 1939 Art worked as a delivery boy for a lunch wagon. Every day he carried hamburgers and soft drinks to the staff of WHDL, the local 250-watter, which was located in a downtown bank building. In this way, he made friends with the station people, and when Art eventually expressed a desire to go on the air himself, the manager hired him. "In one day I went from a job paying $13 a week plus meals," says Art Hannas, "to a job paying $13 a week without meals."

Almost everyone has a contact of some sort. Somewhere there must be a relative, or friend of a relative, or friend working in some phase of the radio-TV business. This is your starting point. It may become a tedious routine of talking to people, experiencing polite turndowns, but in the process getting other referrals. Every lead, no matter how obscure, must be pursued with enthusiasm. As an ambitious neophyte, you should devote the equivalent of an 8-hour day solely to the project of soliciting interviews, making as many contacts as possible everywhere.

On the heels of routine interviews with station personnel, start calling on allied enterprises—the cable companies, advertising agencies, program packagers. Try all of the possibilities you can think of or find in the Yellow Pages.

Salesmen in all fields learn early in the game that a certain number of "blind calls" will produce a consistent percentage of leads over a long period. In time a definite pattern of "yes" replies will emerge. In your favor is the fact that the people in broadcasting tend to help each other. They pass along helpful tips about jobs; not to the extent of cutting their own throats, you understand, but they will tell of an opening that you may be suited to fill.

Such an early contact hit the jackpot for Jack Webb, who was to become an industry legend as producer and star of DRAGNET and later executive producer of ADAM-12, THE D.A., EMERGENCY, and O'HARA, UNITED STATES TREASURY. Everything started for Jack about 1945, the time of the United Nations conference in San Francisco. He met a radio producer who told him that KGO needed radio announcers. Jack had never announced anything, but he went to the station where a woman auditioner gave him a script, pointed to a microphone and control board and said: "Just push the button, young man, and read what's on the paper." Jack Webb read so well that he was accepted as a staff announcer.

All professionals have great sympathy for mike and-or camera hopefuls, and that's another peg to hang your hopes on.

ICE BREAKERS

You can think your way into broadcasting by dreaming up an original program and selling it to a radio or television station—a new kind of game show, informative newsy program, how-to-do-it yourself feature or variety entertainment.

By looking critically at a listing of programs already in existence, you might well say to yourself: "There can't be anything new. Everything that can be done has already been done." Or, "If my program idea were worth doing, someone else would have done it by now." Wrong! Just turn your thoughts another way, and you'll realize that regardless of the number of radio and television programs presently on the air, just as many remain to be developed. Due to the fact that things are always changing there is always a real opportunity for something different.

To get a salable idea, however, requires more than just wishing for it. To acquire the art you must:

Stimulate your imagination

Turn your mind to work on something that truly interests you

Follow a definite process to make it produce.

Most program formats are basically the reassembling and rearranging of old elements into new combinations.

Take radio traffic reporting. In urban areas, announcers like San Francisco's Lu Hurley in his KGO Radio "Hurleybird" flys over metropolitan cities during traffic hours and tells people in their cars what terrible snarls they're getting into. There's nothing particularly original about that type of broadcasting anymore, is there? But, hold on a moment! Kelly Lang, an aspiring young actress, approached KABC, Los Angeles and literally got her career off the ground by persuading station bigwigs that they should put her at the helicopter microphone to sooth the jangled nerves of frustrated drivers below. Kelly then soared to great heights in her silver jump suit and crash helmet as the nation's first airborne traffic girl. Her job-producing idea was simply to apply a new twist to an old concept.

Not so long ago someone realized that deaf viewers, representing a sizable segment of our population, were totally neglected on the television medium. This resulted in the development of news segments specifically programmed for the hard of hearing at various stations across the country. Thanks to this brainstorm, deaf viewers can now get their daily dose of crisis like the rest of us through sign language on programs like KRON-TV's Newsign 4. Here was an idea that had not been explored before, and it performs a definite public service.

A fresh approach to get on the air can be one of many forms. The main thing is having something a bit different to offer and then following a positive plan of action.

Like many hopefuls, Dick Cavett found it almost impossible to crash into the exclusive inner sanctum of network programming. After mind doodling for a time, Dick devised a plan. In 1960 Dick, at that time a copy boy for Time magazine, went to the NBC-TV studios with an unsolicited collection of jokes that he had written for Jack Paar. Upon encountering Paar in a hallway, he thrust an envelope into the star's hands, mumbling something about a monologue. Then Dick waited apprehensively. Two weeks later, after Paar used some of his material on his program and got the hoped-for laughs, Dick Cavett was invited to forsake his $50-a-week job and to begin work as a $360-a-week writer for the TONIGHT show.

Good ideas are around us all the time, disguised as the obvious and the commonplace. And the best way to develop a new radio or television program is to increase your curiosity. Make it the curiosity of a little child, or a foreign visitor just off the pickle boat, or of a man from Mars. Watch for new developments that can be applied to a new program. Read the newspapers and magazines, looking not only at the news and sports columns, but at the ads and the feature articles. Ask yourself constantly: What program idea can I get from this? What can I do to carry it another step?

The simplest form of program is the interview show, which is either good or bad, depending on the personalities involved and the subject matter. Can you think of a new kind of

Hokus Pocus, the popular magician-clown, is a favorite with small fry audiences on his daily shows presented on cable television.

question-and-answer gimmick that hasn't been done? Put your mind to work on this problem.

Suppose you could bring before the cameras week after week con men, burglars and other professional thieves, wearing masks to hide their identities, revealing how they go about stealing from the public. Such a show would capture an audience, wouldn't it? You bet it would, especially all the police in town. You see how you have exercised your imagination, and how it has resulted in something out of the ordinary?

Consider other program formats and put your imagination into action, sifting and sorting angles to give them a fresh, new appearance.

Body building. Because both men and women wish to be healthfully slim, there is a tremendous interest on the part of both sexes to accomplish this. They can be shown just how to go about it through television. Perhaps a great show with a new "hook" in it can be developed by you along these lines.

Travel. Informative programs picturing how people live in other parts of the world are generally popular. With TV, a viewer can look at the world without moving out of his favorite chair. Consider new formats of taking this lazy viewer on trips from the poles to the tropics.

Sports. Every form of sport can be effectively illustrated on the visual medium. Pro golfers can reveal proper grips, swings, stance and follow through. Billiards is pictorial, and ping pong offers action. All sorts of competitive matches are interesting if they are done well. Who knows, soon you may bring electronic checkers or 3-dimension chess or some other indoor sport or game to the video screen?

Arts and crafts. Television, fulfilling obligations of public service, must include the finer things in its program schedule, such as presentations of painting and sculpture, lessons in home decorating, teaching music and dancing, how to make things. Numerous sources of material are available for these features—government reports, university bulletins, standard reference books, private business booklets. An article in a magazine or newspaper could well trigger an exciting new topic for local application.

Kids Shows. Programs designed to appeal to youngsters are always clamored for. Performers work either in regular street clothes or appear in costume and makeup for character roles to excite the tiny tykes. Examples of the latter may be a space traveler like Captain Zig Zag, a circus clown like Hokus Pokus, a cowboy like Ranger Roger or a magic story lady like Princess Pat. For older children there are more advanced programs—new inventions in jets, rocketry, communication. Someone will someday put across a show for kids with dramatic sparkle that will top even the popularity of SESAME STREET. You could be the one to do it.

Variety Shows. A program that takes in the whole gamut of songs, dances, dramatic readings, stand-up comic bits, juggling, snake charming, sword swallowing, etc., is considered broadly under the heading of "variety." A show of this nature can introduce on the screen either professional or amateur talent. In every town in the world are performers eager to present their skills to the viewing audience. Some are retired nightclub and circus professionals; others are youngsters just hoping to break in. Put on your thinking cap and explore the possibilities of adapting the old ED SULLIVAN SHOW or ORIGINAL AMATEUR HOUR for strictly local presentation. People who are talented in any way, even if it's the playing of "The Stars And Stripes Forever" with spoons and forks, are good video material.

Quiz and Audience Participation Program. The development of a giveaway or game show has plenty of potential, and to be 100 percent successful the home audience should participate. An ancient children's game, as simple as the long-running TRUTH OR CONSEQUENCES, packs visual interest. You may locate the clue to a productive show of this kind in a book of games as close as your library shelf.

Modern radio offers a different sort of challenge to your thinking machinery. For the most part, today's radio program directors prefer short features that can be programmed as bridges between phonograph records. These featurettes are usually one to two-and-a-half minutes in length so as not to bore the restless AM or FM listener. A DEAR ABBY advice bit is typical. KNBR schedules THE CALIFORNIA STORY, making good use of historical vignettes. Weekend MONITOR on NBC, heard transcontinentally, is chock full of informative short reports, interviews and comedy stints.

On your hometown station, the chances for you to create and produce short audio subjects are excellent. Station operators will welcome fascinating featurettes that are local in character and which they lack the staff to produce themselves:

Profiles of important local people.

Good buying tips compiled from various stores, food outlets and government sources.

173

A "Help Wanted" show highlighting job opportunities.

"Classified Ads" of the air.

Fashion features.

An "Action Line" featurette replying to letters of complaint from listeners pertaining to bad business practices and city government.

Discussions on diet and health.

Gardening advice.

Movie and book reviews.

Quickie "man-on-the-street" interviews based on questions of topical interest.

Talks directed deliberately at women's homemaking and social interests.

Hunting, fishing and ski information.

An enterprising ex-radio jock and nightclub singer has built a thriving business selling singing commercials door-to-door to the nation's small merchants. The jingle maker and his wife roll around the country in a motor coach that has been converted into a combination recording studio, office and home.

With the big unit parked outside a place of business, the traveling jinglist lures the customer away from his desk, phone and other distractions. Inside the coach, the advertiser hears on 8-track stereo a customized jingle for his business that has been created on the spot. How about a happy singing commercial for a mortuary? Funerals don't have to be dreary. Or something with a rock beat for a furniture store's semi-annual clearance sale? The musical commercial maker has them all on tape—right there in the parking lot.

The bright idea for the mobile jingle mill came into being when the former disc jockey decided there was a large chunk of American business in small and medium-sized towns which never had known the joys of radio jingles. It was a thought that now pays well.

Throw the switch of your imagination to the "on" position. Begin to adapt ideas that will serve your purpose in creating a new kind of show or salable device that will be your ticket to a broadcasting career. Mind you, there is no need to re-invent the wheel. For your purposes you need only improve on it a bit by adding a tire and axle.

PLAY IT COOL IN SUMMER

In summer the world is divided into two classes of people—those who can run away to the mountains or beaches, and those who have to stay and face it. The general situation can be summarized this way: **Whereas winter brings absence of leaves, summer brings leaves of absence.**

This is your cue, job seeker, to take action during the hot months. Opportunities for new people to break into broadcasting are especially excellent in July and August because so many regular staff members go on vacations, and stations need people to man their vacant posts.

Perspiring station managers in wilted collars are not as demanding of applicants during summer as in other seasons of the year. When the temperatures soar to the 90s, these men are more interested in a cold shower or a cold beer than in sweltering over letters of application and sweating out auditions. With a wave of a moist hand, a station manager sometimes directs a newcomer to the control room, and professional experience be damned, if someone is urgently wanted to replace a vacationing DJ.

Temporary summer jobs at radio stations many times turn into permanent positions. Numerous are the staff people, supposedly on holiday, who are reality seeking other jobs during their week or two out of town. If successful in their quests, they never return. The "summer replacement" position then becomes a permanent one for the fill-in jock if he has done a reasonably acceptable job of announcing.

My friend Loren Miller, a talented writer and announcer, once took a summer job at WQBC, Vicksburg, Mississippi. Thirty-six years later he still had it. Poetically expressed:

> Summer enhances
> Your on-air chances!

THE BABY SITTER

Radio station operators sometimes hire part-time "baby sitters" to watch over their transmitters in keeping with FCC Rules and Regulations. These sitters generally come from the ranks of radio amateurs and hardware-oriented electronic technicians. The key to their employment lies in the fact that they all hold first class radiotelephone operator licenses.

A vignette from my personal history typifies what happens at many stations. In the early 1960s I put on the air Radio Station KAPY ("Cappy") in Port Angeles, Washington, operating with 1000 watts of power at 1290 kHz. KAPY was licensed by the FCC to broadcast with a directional antenna, which meant that our signal was to concentrate in certain areas and be kept out of others. The conditions of our broadcast license plainly stated that a first-class operator must be on duty at the transmitter at all times. Therefore, as general manager of the station, I hired only combination men, specifically disc jockeys who held those all-important first phones.

But there was a complication. Because of a limited operating budget and the humane fact that announcers have to have at least one day off each week, I found that we lacked a qualified first phone man to fill in on Saturday and Sunday afternoons.

My search for a part-time radio engineer began. I checked with engineering personnel at the local telephone company and at radio service shops; I placed a want ad in the Port Angeles newspaper; and I inquired around town for a first class radio operator. In a city the size of Port Angeles, population 30,022, I was not able to locate a qualified technician.

As I remember it, I was sitting in my office at KAPY with my feet on the desk, hands laced behind my head, staring at the ceiling and puzzling over how to solve the problem, when a Jaguar convertible containing a dapper, mustached little man of about 50 and two beautiful Dalmation dogs roared into our parking lot. Into the station marched Jim Moulten, who identified himself as a retired Coast Guard chief radio operator, the holder of a first ticket, and a fellow who was interested in filling the weekend "baby sitter" position.

"Money isn't what I am really looking for," he said. "What I want is something to occupy my time." To say the least, I grabbed this remarkable stranger like a long-lost brother who had at last struck it rich in the Texas oil fields and had returned to split his fortune with me.

"Jim," I cried. "Have a cigar! Have a cup of coffee! How about a couple of filet mignons for your Dalmations? Name it, it's yours!" And so Jim took over the controls on weekends, keeping the logs and keeping us legal with the FCC.

Now, take my word for it, you could expect by the law of averages that the one man out of 30,000 to apply for the weekend shift would lack much in announcing abilities. Not Jim Moulten. His voice was deep, his reading technique smooth, and his style on the air not unlike that of Boake Carter, a powerful radio personality of the 30s. In short, as a fill-in combo man, Mr. Moulten for Radio KAPY was solid gold.

By that example, I hope I have driven home how valuable an asset a first class radiotelephone license can be in opening doors at radio stations. Bear in mind that sometimes the transmitter and studio of a station are separated by as much as 15 miles, in which case a first phone man is required to baby-sit the transmitter. To save duplicating salaries, the manager will likely assign the transmitter engineer an air shift. The inevitable conclusion is that the prospective announcer is in a good position when seeking employment if he has a first ticket. A good way to get started.

A COMMERCIAL FOR YOURSELF

Your own resume can be the most important commercial you ever read. This summary about yourself is a sales-promotion piece, and your first job-getting tool. Alone your resume will not get you a job per se, but it will open an employer's door for an interview. The vital information in your summation should be designed to:

(a) Create a favorable image of you in the minds of the people who can hire you;
(b) Remind those you have already seen that you still exist; and
(c) Sell you when you are not present to sell yourself.

Your resume, typed neatly, with a good, clear photo of yourself attached, must present the honest facts about you. It should outline your education, abilities, talents, specialized training, references and other personal data. Without coloring anything about your past history, you must be sure to include in your resume a line or two emphasizing how you can be an asset to the station.

If you lack experience, admit it, but go on to say that you are ambitious and anxious to acquire as much professional know-how as possible.

Think carefully about personal references, and submit only the names of those persons who know you well and whose characters are above reproach. Be sure to cite bank references, and mention any credit cards that you hold from companies, department stores or major credit dispensers such as American Express or Master Charge. Established credit stresses your character as a "solid citizen."

Not only should your resume be typewritten when submitted to a radio station executive, **it should be recorded**. After all, you are dealing with an electronic medium, and sound is its stock in trade. On tape you will want to give your name, address and telephone number, a brief explanation of what you have been doing for the past several years and some concrete reasons why you wish to be a disc jockey or other performer. This gives a prospective employer an opportunity to hear exactly what you will sound like on the air. It also demonstrates to him that you are very serious about entering the broadcasting business.

Keep your resume brief, about one minute's reading time. On paper your resume should break down categorically as follows:

RESUME

Bill Holiday
721 Downy Drive
Goose Falls, Nebraska 12345
(711) 923-9829

Age 23

Job objective: To become a disc jockey at Radio Station KZAM, and to learn continuity writing and selling on my own time.

Professional training and special talents: Radio announcing home study course. Have acquired third class radiotelephone operator's license. Participated in high school dramatics, Armed Forces Radio Service, and emceed service camp broadcast shows.

Education: Graduate of Goose Falls High. Took courses in public speaking.

Employment: For two years bellman at the Feathers Hotel, Goose Falls.

Personal references:

    Robert J. Fosdick, Treasurer,
    First National Bank.

Harvey Cromfeller, Principal,
Goose Falls High School.

Mrs. Dawn Renton, Manager,
Feathers Hotel.

Credit references:

Ping Oil Company Credit Card.

Beak Clothiers, 210 Main St.,
Goose Falls.

Marital status: Married, one child.

Military: Honorable discharge, U.S. Army. No reserve obligation.

On tape your resume should read much like a commercial:

"This is Bill Holiday of 721 Downy Drive, Goose Falls, Nebraska. I am 23 years of age and married with one child. My ambition is to come to work for Radio KZAM as an announcer and to learn continuity writing and selling on my own time.

"Although I have not had any professional radio experience, I have prepared myself through home study to become a disc jockey. I have acquired a third class radiotelephone license in planning my broadcast career.

"At Goose Falls High, the school from which I was graduated several years ago, I participated in dramatics and took courses in public speaking. While in the Army, fulfilling my military obligation, I announced some programs for Armed Forces Radio and I emceed several amateur camp shows.

"I am at present employed as a bellman at the Hotel Feathers, in Goose Falls, but I feel that my real future belongs in commercial radio and that I have something special to offer. On my written resume, submitted with this tape, are personal and credit references and other pertinent information.

"Thank you."

THE WRONG LINE

In trying to set up an interview or audition at a broadcasting station, the lazy aspirant's first thought is to use the telephone. Don't allow yourself to fall into this dangerous trap. **Always appear in person!**

From your own experience, you know how easy it is to brush off someone on the telephone. A sales pitch to sell you a set of encyclopedias made on the phone, for example, stands as little chance of getting a hearing as a cough in a whirlwind. On the other hand, if your doorbell rings, and a friendly person stands outside your domicile with an armful of books, shooing him away is not as easy. The same principle applies to the busy station executive. If you should reach him at the wrong moment, and he hangs up on you, you have blown your chances.

Occasionally, job seekers use tricks and gimmicks in their efforts for personal promotion. Sometimes a photo is submitted in the form of a jigsaw puzzle, or a cigar is sent to a program director with a resume attached. One hopeful put an audition tape between two slices of bread and mailed it to a station manager. None of these angles have worked that I know of. Hunting for work is a business, and you must approach it in a businesslike way.

THE INTERVIEW: NODS AND NO-NO'S

Comes now your interview at a broadcasting station. Keep in mind that all interviews and auditions have one thing in common: the necessity that you favorably impress some one person or group of people who may hire you. And remember that all prospective employers want to know **what you can do for them.**

Be aware at all times that you are trying to market a commodity in a buyer's market. There are more beginners than there are jobs. The broadcasting industry claims that it is constantly in search of new talent, and while this is true to a certain extent, the fact remains that talent is usually in search of broadcasting. Your main objective is to sell your talent in such a way that it will be remembered. Since the first impression you make is your most important calling card, be sure it is not thrown in the wastebasket.

Be punctual. Make it a point to arrive at the station on time. Being late would be unpardonable. Since broadcasting runs on the second-hand of the clock, so should you.

Dress conservatively. A man should wear a suit and tie; and a woman, a simple dress.

No booze. Don't stop on the way to your appointment for a bracer of any sort. The telltale alcohol clinging to your breath could puncture your opportunity.

The first person that an aspiring DJ meets at a radio station is the receptionist. A favorable impression made in the lobby with her can help open many doors.

Be nice to everyone. Try to make friends with everyone you meet in the studios and offices; it can do you a lot of good.

Don't smoke, unless the interviewer does. A cigarette lighted in the presence of someone who disapproves of tobacco can be dangerous to your chances.

Avoid "yeahs" and "uh huhs" in your conversation. Impress the man with "yes, sir" and "no, sir" which are heard too little in today's arrogant society.

Be humble. Don't assume that you know more than the person who can hire you.

Don't monopolize the conversation. Try to get the interviewer to do most of the talking by asking him sensible questions about his station (based on your research of it). Above all, an employer wants people who **care** about working for him.

Show your genuine enthusiasm. Job opportunities are often lost because of seeming indifference on the part of the applicant.

Never ask about fringe benefits. When the employer has decided he wants you, he will begin to **sell you** on taking the position and will discuss benefits at that time.

Don't be a name dropper. Mentioning that you have dated the station owner's daughter or that the mayor of the town is your first cousin, won't help you a whit. It will probably antagonize the interviewer.

Girl hopefuls should not try to turn on sex. Despite all that has been written and discussed, very few jobs in show business are gotten on the "casting couch."

Be relaxed. Be yourself. Accept criticism cheerfully. One of the hardest things to quarrel with is a friendly smile.

Don't for an instant doubt yourself or your ability. Think positively that you are going to succeed in getting the job.

Following the interview, mail a "thank you" note to the interviewer the same day. In it, review highlights of the meeting and bring up additional points that may have been overlooked.

From my own happy experience, I can illustrate how very important this follow-up can be. Years ago, just prior to putting Radio Station KAPY on the air, I advertised in the Port Angeles, Washington, newspaper for a Girl Friday. More than 70 applicants applied. From this number I reduced the list of possibilities to three.

While contemplating which of these three to hire for the position, the mail arrived. There was a "thank you" note from Miss Karol Newlun, one of the three young ladies I had in mind. Her handwritten letter got her the job. And also myself. A few months later we were married. That happened more

KXRX Station Manager Charles "Chuck" Cristensen interviews a job applicant.

than a dozen years ago, and I consider Karol's little letter the finest literature ever written.

ABCs OF AUDITIONING

(a) Take an audition at a broadcasting station only when you are ready for it. This is your very real chance for an announcing job, so don't put your product on the market until it is completely off the assembly line.

(b) Don't overdo your audition; make it simple. By doing what you do best, and keeping the presentation brief, you give the auditioner something to remember you by.

(c) Be business like, and not awed by anything or anyone. Treat your audition as you would treat any other employment application. Bear in mind that the station auditioner is as anxious for you to succeed as you are to be successful.

IF AT FIRST YOU DON'T GET IN, TRY, TRY AGAIN...

Breaking into the broadcasting business may be discouraging at first. If you are easily disappointed, or if setbacks weaken your resolution, then in all fairness, radio and television are not for you. Getting started on the air requires determination.

You may hear the words, "No, not a thing" many times. But if you accept the challenge that **a door will open**, those disenchanting words will after awhile lose their meaning.

Think optimistically, and resist the temptation to slow down. Don't get in the position where you begin to think of interviews as a waste of time. In the profession you have chosen, you just don't know where or how a job will turn up. **But it will!**

Closing Commercials

PART 4

BROADCASTING AS A BUSINESS

To the owners of commercial broadcasting stations, money is not the root of all evil. The lack of it is. Money ranks as the chief reason why station operators entice talented talkers to their microphones and cameras. The formula is simple: In any area the most popular radio and television stations get more money for each commercial minute that they sell, and thus enjoy larger profits. Naturally, then, it pays to own the most popular station in town.

An advertiser's interest in broadcasting is, basically, to obtain circulation at a rate lower than he can obtain it in the printed media. And this means people. People listening to radio. People watching television.

To achieve popularity and to ring their cash registers with fat advertising dollars, commercial stations are always on the lookout for talented people who will appeal to the general public and build larger audiences.

The name artists in major cities earn impressive salaries and talent fees because they are particularly popular with audiences—John Gambling in New York, Wally Phillips in Chicago, George Putnam in Los Angeles, Hal Lewis ("J. Akuhead Pupule") in Honolulu.

In the Greater San Francisco Bay Area, Russ Coglin is typical of the all-around professional who has come to the top in radio and TV by imagination, resourcefulness and his keen ability to sell his audiences. Russ' personal endorsement of a product stimulates the masses to buy it. That's broadcasting salesmanship at its best, and a performance quality aggressively sought after by commercial stations.

Like Russ, the performer who has something special to offer is a much-wanted man. And whether or not a station has an opening for a performer, when a really outstanding personality presents himself, they make room for him. This principle scales all the way down from top network stardom to the disc jockey who is witty and bright in the little burg. As a result, the manager of each station in any city is actively seeking on-air people who are interesting sounding and who have "sell" in their delivery.

Therefore, in your professional career you will have to be two things:

1. Actor.

2. Salesman.

As an actor you must create different moods with your voice to fit the material you are handling. Sometimes you will be friendly and informal. At other times, you will be strong and dramatic. **But, first and foremost, you must be a salesman!** While voicing commercials, you have to tailor your voice, your inflections and everything about your presentation to get that important sales message across in a convincing manner.

The radio disc jockey is primarily a merchandiser. In the final analysis, he is an announcer with a special aptitude for pleasant and interesting chatter in between the playing of records. He cleverly injects his own personality into the framework of a program and puts salesmanship into what he says. Henry Morgan, Murray the K and Jim Lange have this unique selling quality.

On television, a "personality" differs from a straight studio announcer for the same reasons. As an average viewer, you are constantly intrigued and perhaps overpowered by pitches by George Fenneman, Jack Clark, Betsy Palmer and Ed Reimers, as opposed to run-of-the-tube spoilers doing commercials for constipation pills, stomach soothers and nasal-passage remedies whose names forever remain unknown to you.

Never forget throughout your career in broadcasting that a commercial must be a pause that impresses. **The better you can sell on the air, the more money you will earn for yourself.**

COLD CASH AND HOT MICROPHONES

You have "arrived" as a disc jockey when the leading restaurant in a big city names a sandwich after you and prints it on the menu. In the meantime, however, you are going to have to select something of a practical nature as your immediate goal.

In general, embryonic announcers make a start at 250-watt or 1000-watt stations in a small city of 5,000 to 30,000 population. These typical operations hire from three to five announcers each, and salaries range from $80 to $150 weekly for 36 to 40 hours.

Many small-time station owners welcome the novice announcer on an earn-as-you-learn basis. The station operator pays only the minimum wage allowed by law. But don't knock it—this is at least a starting point! Looking on the bright side, the beginning DJ benefits from technical instruction and gathers practical experience through on-the-job training. To survive on pitifully small paychecks, the jock-in-training has to tighten his belt and make do until the trumpets of success beckon.

In the beginning, at meager wages, you will have a mind-blowing job. You'll be on the air up to six hours a day, performing all the tasks connected with broadcasting. You will no doubt work in a small control room, pulling your own records, compiling news and sports, keeping station logs and, in short, doing everything that needs doing.

I remember my first announcing job after my discharge from the Navy at the end of World War II. Sitting cramped in a tiny, cluttered control room with a pot of coffee and sandwiches at my elbow, I pumped out music and commercials for ten straight hours, while the owner spent his time on the street selling more commercials for me to read. Throughout the long day I saw not one human face. Yet I did have companionship. It was Caesar, the owner's colossal, old and graying German Shepherd left behind for me to babysit. Caesar had the bad habit of howling most everytime I opened the microphone, and the listeners in radioland must have thought I was beseiged by wolves. The job paid $1 an hour.

Getting back to your own guest for broadcast employment, I must emphasize that if you are a fierce advocate of the 4-day work week, small-time radio is going to be a disappointment. You can expect to put out six days of every seven. Since radio never takes a day off, you should consider yourself lucky, perhaps, to enjoy one.

But as time goes on, and you gain more and more experience and the resulting self-confidence, you will move up from the small, very local station to a larger one where more emphasis is placed upon specialties, such as news, sports, interviewing or jockeying records. In this decidedly better position, you will also get help in programming, engineering and two days off each week, not to mention a huskier paycheck. In larger cities, salaries may range from $10,000 to $25,000 a year and more. When announcers reach this level they often develop their programs on a contract basis with a station which pays them a percentage of the income their program makes from commercials, and many such announcers earn well over $100,000 a year. In addition to these

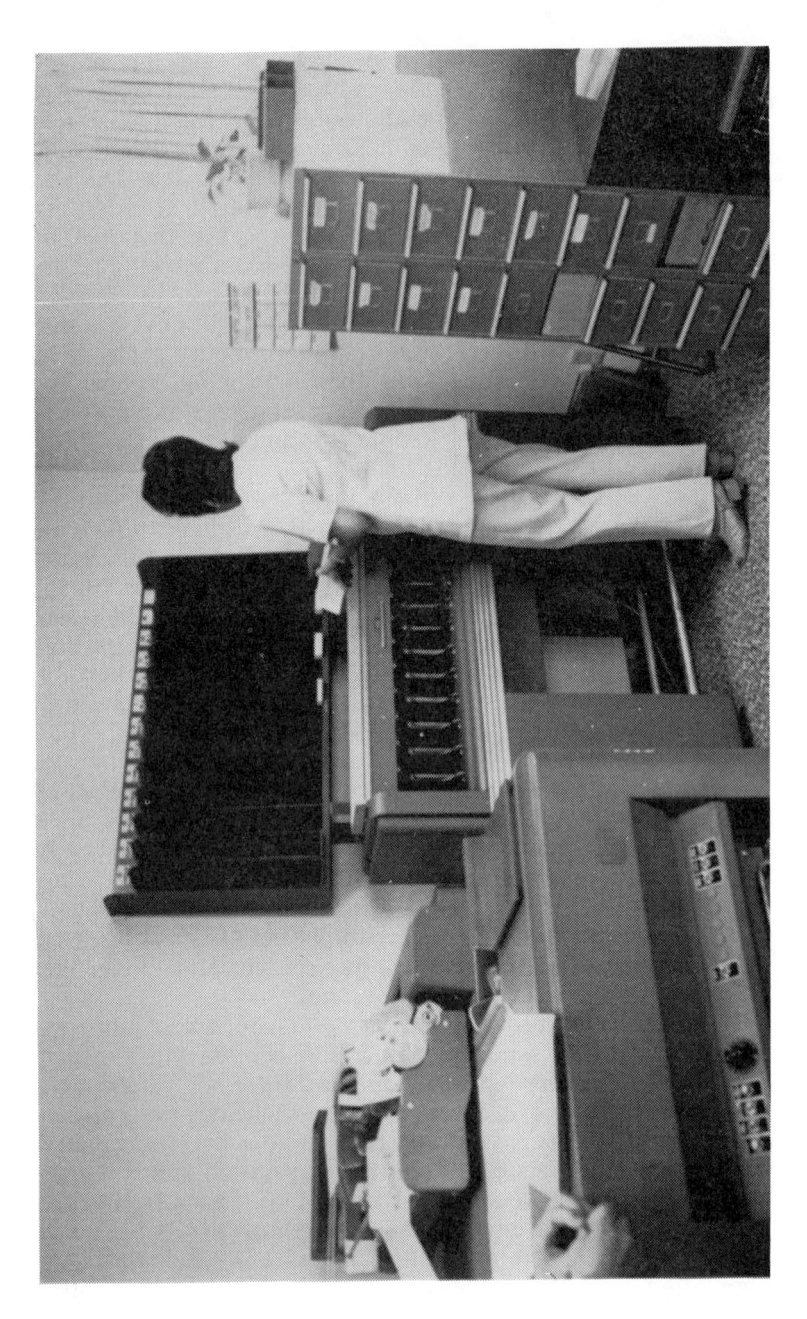

There are opportunities for women in the traffic department of a broadcasting station. A traffic girl maintains files of all formats, commercial copy and books sales orders. She also keeps the talent mail count.

financial possibilities, radio and television announcers often work a part-time day. It is not uncommon for an announcer to work only two or four hours a day.

If you stick to your guns and enjoy a few breaks along the way, someday you could become a top-notch DJ, making a quarter of a million dollars a year or more, with secretaries, writers, program assistants and technicians to take care of much of the labor that you have done all by yourself as a small-station jock.

Frankly, the job outlook in broadcasting is cheerful, despite the grumblings of some pessimists who mutter that it is extremely difficult for beginners to find work. Just sour grapes. Close to 100,000 **full-time** artists and craftsmen and approximately 25,000 **part-time** workers are employed in broadcasting. Statistics reveal that about 65,000 of these are in radio, and that the annual turnover is about 33 percent. These are welcome figures for a would-be jock. And there is additional good news: More than 2000 radio stations in the United States are located in cities of 10,000 souls or under, exactly the size of market you would logically look first for an announcing job.

Performers on the way up usually join the union army, specifically the American Federation of Television and Radio Artists (AFTRA). This is a collective bargaining agent for wages and working conditions which represents announcers, actors, narrators, newscasters, commentators, sportscasters, masters-of-ceremonies, moderators, quiz masters and others. AFTRA sets all the rules, regulations and wage controls for its membership, and is designed to protect its radio and TV performer members from any unfairness.

Scale fees are established for station staff members and for free-lance performers, including residual (re-use) payments for all commercial work. I have always found that reading the AFTRA payscale books, with rates that vary from city to city, as difficult to decipher as the long form of the Internal Revenue Service. But, as a general example, I can tell you that in major cities a voice over announcer of a commercial minute earns around $100 for its first 13 weeks of telecast. Close enough?

But first things first. If I were you, I wouldn't fret about union scales and big money at the outset. I'd concentrate on my first paycheck, small though it may be, earned from broadcasting. Starting at the bottom of things is usually what lands a man on top eventually. Think like the noted philosopher who sagely commented: "Gentlemen prefer blondes, but will take anything."

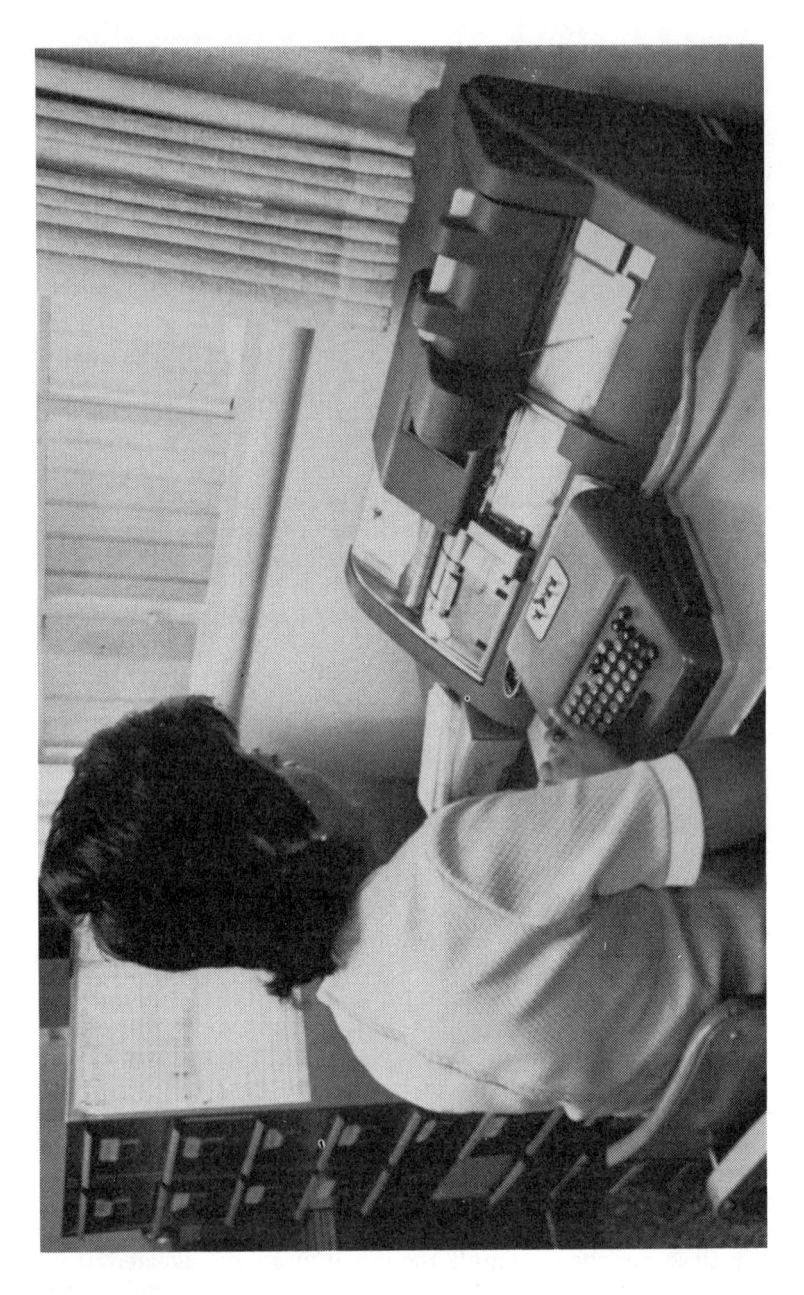

Computers play an important part in the operation of many modern radio stations for logging, billing and the selection of musical numbers. In this shot a traffic girl is shown feeding the machine information for tomorrow's log.

AUDIENCE REACTION

The important thing is not that money talks, but that it has the largest listening audience.

POWDER POWER

Never, never, never underestimate a woman in broadcasting, unless you are speaking about her weight or her age. The day has dawned at last in the broadcast world when it doesn't matter to an audience whether facts and opinions, ideas and insights are communicated by a man or a woman, or whether a woman is young or not young, or whether she is a former Miss America or a Playboy centerfold. The gals have arrived to stand on their own nyloned legs and dish it out professionally.

On the scene going great guns are Pauline Frederick, reporting to us from the United Nations; Marlene Sanders and Liz Trotta covering the latest complications in East Asia; Betty White describing the action at a special event; Judith Crist lambasting or lauding the newest cinematic attractions. These ladies are superb professionals who have a special way of looking at things with a kind of intimacy of vision.

In the art of interviewing, a woman gets a different set of answers than a man. For example, a battery of men reporters questioning a public figure will ask if he believes such and such to be correct, and what about his statements concerning so and so. Then a gal reporter steps up and asks where he spends his spare time, what he eats for breakfast, and how he communicates with his teenage son. A woman questioner very often turns what would otherwise be a dull interview into something that's unexpectedly entertaining. What's more, I have to say (at the risk of being kicked out of the Society For The Continuation Of Men As Lord And Masters) that in many instances a woman announcer delivering a personal endorsement for a product somehow packs more sales wallop than a testimonial from her male counterpart.

Radio disc jockeying continues as an almost all-male occupation. Although a substantial number of women commentators are employed to handle homemaking and community-interest programs, very few women staff announcers can be heard. The irregular hours of work and the necessity for operating technical equipment are reasons given for the scarcity of women as DJs, but privately a majority of radio executives feel that on the air women, as a general rule, are "too patronizing."

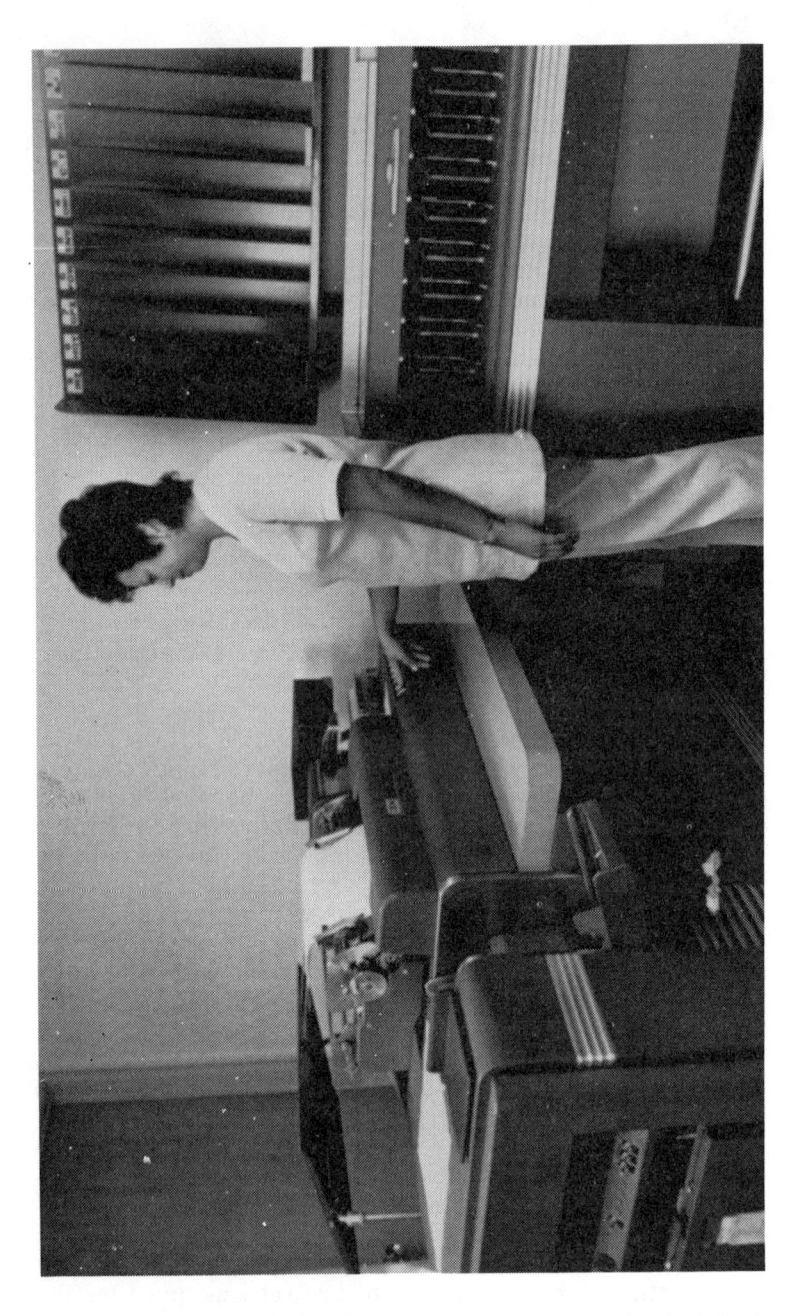

Together computers and the Fair Sex play important roles in today's modern broadcasting stations.

On television it is a different story. The opportunities for the girls as hostesses, demonstrators, interviewers and program announcers are definitely on the increase. Any list of established performers should include such names as Carol Burnett, Dinah Shore, Margaret Mead, Julia Child, Barbara Hale, Bess Meyerson, Virginia Graham and Barbara Walters. The list grows longer every month. Time was when weather forecasts were assigned almost exclusively to men. Now many women handle them, and well. This trend is rapidly expanding to hard news reporting, and, occasionally, to sports.

Of considerable help to women in broadcasting was the Federal Equal Pay Act of 1963, which states that: "The federal law prohibits employers from discriminating on the basis of sex in the payment of wages for equal work on jobs requiring equal skill, efforts, and responsibility and which are performed under similar working conditions."

One important nonprofit, professional organization encouraging increased female participation in broadcasting is American Women In Radio and Television, Inc. (AWRT) with headquarters at 1321 Connecticut Avenue, N.W., Washington, D.C. 20036. Feminine readers who are interested in receiving free literature on careers for women in broadcasting should write to AWRT, Inc. at that address. Whoever said that "women should only be seen and not heard" is fast losing ground to the gals who are coming into their own in broadcasting.

GOOD RECEPTION

No one knows like a woman how to say things which are at once gentle and deep.

Victor Hugo

THE SKY'S THE LIMIT

In New Orleans not long ago, I visited an old friend, Douglas Ellison, Vice President and General Manager of WTVU-TV, the ABC station for the Southern city. Sitting in Doug's impressive office, sipping Channel 8s biting chicory coffee, we reminisced about old times, and it crossed my mind again how broadcasters like Doug, with ambition and dedication, can start from humble positions in the media and progress steadily upward to head important outlets, program packaging companies, and sometimes buy their own radio and television stations.

On a live program, weather girl Claudia Nygaard forecasts snow for the Great Lakes region.

Doug Ellison recalls, for instance, how he originally struggled as a classified ad salesman for the San Francisco Chronicle in the late 40s, and how he switched over to KRON-TV as an announcer-director when that newspaper-owned station first went on the air. At a beginner's salary of $75 a week, Doug got going in the visual medium.

Prominent broadcasters at the top of the field are often asked what the secret of their success has been. A question like this implies that there is some secret plan, a magic ingredient that guarantees prosperity to anyone who knows about it and uses it. The hard fact is that, there is no one "secret" or thing that insures success for everyone. Just check the biographies of many famous radio-TV personalities, and you will find there are varying answers to the question.

Paul Harvey, for more than a quarter of a century, an ABC commentator and a syndicated columnist in hundreds of newspapers, at age 15 was helping support his mother by announcing for Radio KVOO in Tulsa. For years BONANZA's Lorne Greene was a radio newscaster in Canada.

After a couple of very lean years in New York, trying to break into show business, Jack Lemmon, a true child of broadcasting, got his first break in 1948 in a radio soap opera. This led to television jobs, and during the next five years Jack appeared in close to 500 programs.

Marlene Sanders, the first woman anchorman on national television news, was originally hired as a production assistant for newsman Mike Wallace. Successively she became an associate producer, a radio news director, and a news correspondent.

Ralph Edwards, the creator-host of THIS IS YOUR LIFE and other important shows, pursued a career in earnest since he was a school-age boy. He sold his first radio script to an Oakland station for $1.00 before he became a staff announcer for CBS Radio.

Orson Welles, Richard Widmark, Tony Randall, Shirley Booth, John McIntyre and Arlene Francis were all performers on the old daytime serials. George Kennedy of SARGE was a child actor on such radio programs as LET'S PRETEND, and in his teens was a disc jockey on a Long Island radio station.

Art Linkletter, who now stands at the very top of the best-loved, highest paid and wealthiest of TV performers, became interested in radio while a student at San Diego State College. Art rose rapidly from announcer on the local station to program manager, and then to free-lance performer, program packager, and on to stardom.

William (CANNON) Conrad went right into radio after graduating from Fullerton Junior College in Southern

California. Later he was heard as Matt Dillon on radio's early version of GUNSMOKE.

Barbara West, Macy's Broadcast Director for Northern California, began her career in Oklahoma City as a radio copywriter, working two hours a day at 90c an hour. Later she produced wrestling programs for television.

Governor Ronald Reagan of California in his early years was known to radio listeners in the Midwest as "Dutch" Reagan, a sports announcer.

If there is any single "secret ingredient" for fantastic success, it may be summed up in two words: get started.

EQUAL OPPORTUNITY EMPLOYMENT

"I grew up with a kid named Harry Reasoner. I don't know whatever happened to him, but I'm doing all right."

Sam In Shorts

FOR THE LOVE OF MIKE AND CAMERA

There used to be a catchy headline in a famous advertisement: "They laughed when I sat down to play." Take my word for it, this line does not apply to anyone who auditions for radio and television work. Broadcasters are in dead earnest about finding enthusiastic new people who have talent. **They want you to be good!**

A few years ago a well-meaning friend proposed Brenda Sykes' name to the producers of ABC's THE DATING GAME. What the producers saw in this 99-pound bundle of beauty and personality, they liked; and it was just a matter of time before the beautiful assistant teacher from UCLA bowed out of teaching in favor of acting. In a short time she had parts on TV's ROOM 222 and MAYBERRY R.F.D., and in major studio movies. I can well imagine, from my own experience, the good feeling that came over the person who gave Brenda her first boost toward fame and fortune.

In the early 50s, during the days that I was a program packager, I was impressed by the talents of entertainer Rusty Draper, a fresh and freckled, red-haired country boy, who performed nightly in Will King's Koffee Kup, a small San Francisco club. Around Rusty's singing and guitar playing, I formated a half-hour variety TV show, which I interested the Tappan Stove Company in sponsoring. Rusty caught on with viewing audiences practically overnight, and the show sold stoves like griddle cakes. Pretty soon Rusty was recording golden hit after hit for Mercury Records, guesting on THE ED SULLIVAN SHOW and other really big ones, and, within several years, he shot upward to international acclaim as a Country-Western star. I like to think back, "I knew him when..." Like myself, every program producer, talent agent and station executive in this business likes nothing better than to bring to public attention a new face or a new voice.

Take Russ Coglin, San Francisco's top radio communicator. Russ used to be program manager for an Oakland radio station, and in that capacity gave some of the now-famous artists their first jobs, which he enjoys remembering. His most notable discovery was that of the incomparable Phyllis Diller.

"Phyllis had read Norman Vincent Peale's THE POWER OF POSITIVE THINKING between housework and caring for her five kids," Russ recalls. "She wanted a chance to write commercials and get something special out of life. I handed her a couple of assignments, and, of course, she came through.

"Rod McKuen was a high school student whom I hired as a janitor for the station and later put on the air Saturday nights spinning discs and reporting high school news. A few weeks later Rod asked if he could perform some of his original stuff on the program, and I said okay. I felt that Rod had a lot going for him, and it turned out I was right."

In citing these case histories, I have tried to nail down an important point: **The person who auditions newcomers wants to find and encourage new talent. He wants to someday look back and say: "I gave that kid his first job. I certainly can pick them."**

Producer Don Fedderson, I am sure, remembers with satisfaction that most unusual day he first noticed the fantastic Liberace entertaining in a bistro, and how he developed the name of the then little-known piano player into a show business legend.

Comedienne Rose Marie no doubt fondly recalls first setting eyes on funny-man Tim Conway on a Cleveland television show, and recommending Tim to Steve Allen to expose his bumbling bit to a coast-to-coast audience.

And Ralph Edwards must be equally proud of his discovery of Bob Barker. Edwards, while searching for a TV host for TRUTH OR CONSEQUENCES, heard Barker on his car radio in frantic freeway traffic. Bob was conducting an interview on a small local station, but Edwards knew in-

This is a typical control room, the very heart of a radio station. Dan Slattery, a veteran combo man, is pictured on duty.

tuitively he was the one to emcee his national show. **Producers like to spot a winner, and to help a newcomer along.** You? Nothing could please us more!

SOMETHING GOING FOR YOU

Some announcers have it. Some do not. On one side of the talent coin are the broadcast artists who are literate and fresh of wit, or those who are folksy and friendly with down-to-earth, likeable personalities.

On the reverse side, we see and hear the uninspiring announcers who seem merely to go through their motions like robots on a wire, leaving us with a blah, so-what attitude. So why the difference between the have-got-its and the have-nots? Is it a quality a person is born with?

Granted, some persons come into this life endowed with a greater talent than others, but, in thinking it over, I believe that the sharpest announcers are those who work very hard to train and study their trade, and keep right on following this routine throughout their careers on the air. They never let up. The so-so types, I am afraid, are those who charge off to glory only to learn they are not prepared, and who are too lazy to develop their skills.

Preparation, without doubt, is Rule One for selling yourself on the air, and it is followed very closely by **attention to detail**. Top-flight performers, with few exceptions, research their material thoroughly before they appear on the program.

Talk show hosts like David Frost, Ralph Story and Owen Spann do their homework well and come to the studios armed with "now" questions for their interviews. On camera their pertinent questions and witty quips are most times the result of burning the midnight oil preparing for their telecasts.

Successful deejays of the calibre of Gary Owens, Dan Sorkin and Jim Lange keep paper and pencil handy so that they can jot down forthcoming "ad libs." In this way, the leading jocks steer away from hackneyed phrases, and put more zip into their delivery.

Yes, the habit of preparation deserves first place on the list of qualities essential to success in the broadcasting profession. Time and time again I have seen young announcers come into a station and wreck their grandest opportunities because they were not prepared to handle the commercial assignments. Probably the most important single thing I have learned from many years of close association with the pros is that, in the long run, it takes less time to be thorough and to do the job right at the outset than it does to dash into something half cocked.

To the DJ: As a radio disc jockey, you must give audiences a reason to tune you in. To start with, your choice of music is of primary inportance whether you are working Top 40, Country and Western, Classical or Middle-Of-The-Road. All musical numbers must blend together, come out of speakers smoothly and fluently. This means programming only those records that the majority of listeners of your station wish to hear and not the personal preference of yourself or your girl friend. Consistence of quality sound is what bags audiences and holds them. **But one bad record can send listeners scurrying to other frequencies, tuning you out.**

To stay on top of the situation, you should select your music carefully, and at least a quarter hour before you start your on-air shift. "In your head" listen beforehand to the show that you are going to present.

Easy-to-listen-to production demands planning and forethought. It means getting to the station ahead of time, checking the log in advance for new features and new commercials, and thinking out what you intend to do during your three to four hours on the air. Read over all the new copy to yourself so you won't have to read it cold once it pops up on the schedule. In other words, assemble all the elements of your show before you sit in the DJ chair so that you can run a "tight board" from start to finish.

Try to find a "hook" for your program, some kind of contest or feature that fits in with station policy, and one that will catch the imagination of the audience. Maybe you will create a fan club and mail out bumper stickers. You might acquire a pet mynah bird as a station mascot whom you talk to on the air. Perhaps you will conduct a talent hunt for the prettiest girl in town (or the one with the best legs) and inspire your listeners to send in photos of their nominees with prizes to entice entries.

Famous recording artists will help you promote your show. You can request top singing stars to tape special messages for you by writing to the companies they record for. This is good publicity for the recording artists, and your listeners will certainly be impressed to hear something like: "This is Dean Martin. Be sure to tune in the Bill Holiday Show right here on Radio KZAM. It's full of gags, gimmicks and grabbers."

In the final analysis, it is you, the DJ, who runs the show. Every instrument is under your command, and through your

The Tele-Prompter is a mechanical device installed on a studio camera. It is controlled by a floor director and operates much on the fashion of a player piano roll. From it a performer reads his lines while looking directly into the lens of the camera. In short, the "prompter" eliminates the necessity of memorizing lines.

own skill and talent you must bring out the best in everything connected with your program.

Showmanship is the technique that accounts for success in any public entertainment, and it is nothing more than the ability to hold the attention of others. There is always room for improvement for everyone, and the jock who wants to do a better job should come to the studio with fresh material to sprinkle into his program. Search the newspapers and magazines for oddities in the news that will interest your listeners. Seek out funny one-liners that will get laughs. Look for human-interest stories that your listeners will enjoy hearing.

There are times when a humorous story, well told, will make a program sparkle. At other moments, a short, barbed epigram is appropriate. Sometimes a witty definition helps. But humor must never be dragged in, or consist merely of a series of irrelevant jokes. It must be to the point, quickly delivered, and without any labored introductions. By all means, familiarize yourself with your material before-hand and get it pat in your mind to such an extent that when you use it on the air it sounds spontaneous and unrehearsed.

A disc jockey who appears to be quick-on-the-trigger with clever ad libs is usually the guy who has prepared his witicisms well in advance.

For the TV announcer: The studio announcer on the visual medium faces one of the toughest jobs in the industry, for he must be welcomed into homes, not once but many, many times. If a viewer turns on his set and sees a personality he does not like, it is obvious that the viewer will change to another channel. Thus, the TV announcer must face the cameras thoroughly prepared to be good.

He must by constant practice learn to read from a teleprompter or cue cards without sounding stilted, and deliver his lines with pacing and accuracy. Most importantly, he must appear at ease and sincere. The ability to move and gesture effectively comes from practice, and lots of it.

The studio announcer's first step in doing a commercial begins with an analysis of his copy. What is the product, and what are the chief sales arguments? Since different sentences and sales points require variations in delivery, he must rehearse his lines repeatedly to know where to effectively punctuate orally.

Remember that when you are accepted to do a commercial, you are of the immediate concern of many people who expend considerable effort and money in your behalf—the sponsor, director, engineers, cameramen, floor manager,

A studio crew sets up a shot for a car commercial.

writers, ad agencies, and others. For this reason, if for no other, you can understand why it is essential that your presentation has been tailored to perfection in practice. Getting started as an on-camera announcer is one thing; to become popular and achieve "staying power" is something else. You will find it necessary to keep a constant inventory on your attributes and your faults. Keep improving yourself!

Improvement means taking what you already are and making it better. By no means does this imply that you should change into something different, for it is most important you be original. But it does mean that you can improve your gestures, movements, facial expressions, diction and general presentation.

Perchance you'll want to grow a mustache or a beard, or shave them off if you are already bewhiskered. You may wish to substitute contact lenses for your eyeglasses. If you are going thin of hair on the scalp, a hair piece may help your appearance, or a complete "rug" if you are bald. Possibly a different style of wardrobe could give you a better look. And what about your teeth? Ought they be capped? Many professionals insist that personality is 90 percent of what makes a TV announcer successful. And the development of such a personality hinges on one important factor: **Preparation!**

PRE-RECORDED

All the world's a stage, and most of us need more rehearsals.

COFFEE BREAK SALES SEMINAR

There are some simple rules that you ought to know about in preparing a sales pitch and prying advertising dollars from local advertisers.

I hear you asking: "What do I care about selling, since I intend to be on-the-air talent exclusively?"

Yours is a good question for which I have equally good answers. First, it is helpful that you, as talent, realize just how much sweat and how many tears go into the selling of a commercial. After you have spent some time on the street matching wits with fussy, and often unreasonable, merchants you will better appreciate the effort that has gone into getting them on the air. Consequently, you will treat each commercial with tender, loving care.

Secondly, selling is a good means of supplementing your income and of making yourself considerably more valuable to the overall operation. In lots of cases, if you are the emcee of a popular show, the station manager will ask you to make direct calls on sponsors with his sales people to either get new business or for reasons of public relations to strengthen the advertiser's confidence in the station. For these reasons, you will want to be as knowledgeable as possible of general sales techniques in order to speak intelligently at such get-togethers.

Selling is a personal thing—different to each person—and the art of selling radio and television schedules cannot be learned from a book. What can be learned, however, is the fundamental approach to the problem, and something about the market-place itself.

Salesmen very seldom sell for their own amusement. They do it for one thing—money. And that money is out in the bowels of a town, and has to be wrenched loose. It follows then that the persons who know the most about the market-place are going to be the most successful.

To my way of thinking, a do-it-yourself radio-TV sales kit is made up of the following:

1. Average I.Q.

2. A salable idea.

3. Enthusiasm.

4. Ballpoint pen and contract forms.

5. Walking shoes.

6. Continuing optimism.

Before charging out of the station to begin pitching potential advertisers, a bit of homework is mandatory. To begin with, you will want to jot down a list of prospects.

A prospect is any business firm, within the coverage area of your station, that sells goods or services to customers. Several prospect sources are:

A. Advertising in newspapers.

B. Handbills.

C. Advertisers on competitive radio or TV stations.

D. Direct mail advertising.

E. Business building permits issues.

Before you leave your office to make your first call, you must plan your campaign carefully. The organization of your sales pitch is even more important than getting your body in front of the prospect. You should never call on any prospective customer until you are prepared to advance at least one sound idea why he should advertise on your station. There is always such a reason, and you can find it.

In preparing your sales talk, you should analyze it critically.

Is it honest? Never promise anything that you cannot deliver.

Will your plan interest a sponsor? Does it contain all the necessary ingredients, or would some small addition or rearrangement impart a greater element of interest to an advertiser? Work over your presentation until it is 100 percent acceptable.

Do you believe in it yourself? If you are not sold on the plan entirely, for one reason or another, it is not a salable idea. Scrap it and forget it! Get busy and devise another plan which you sincerely believe will produce results for an advertiser.

Let's say that you are going to call on a motorcycle dealer. You might devise a sales-promotion plan in which you agree to ride one of his cycles around town and personally endorse his product on the air. You could further develop a contest in which listeners guess how many miles to the nearest tenth of a mile you will get from 100 gallons of gasoline. Such a gimmick would emphasize the economy of the advertised product.

In all cases, you must create in the prospect's mind a clear picture of how he is going to gain by advertising with you. **Think always of the advantages to the advertiser, and how he can increase his profits.** Only when your prospect is convinced that broadcast spots will make money for him, will he part with his cash.

Confidence in what you are selling is a valuable quality. A prospect senses such confidence and will usually listen with a keen interest, believing that what you have to say is important.

Subtle flattery will benefit you. A good salesman implies in his attitude and his voice that the prospect is a man of high intelligence. He tries to create the impression that they are two people discussing a merchandising plan which can benefit both. Quoting the prospect's own words is a powerful sales device. This sort of playback is flattering, attention-getting and effective, because it shows him how much you appreciate his thinking.

It is wise to agree with a prospect as often as possible. He likes to be right. If he attacks your idea, you can counteract his arguments by saying, "Yes, I do see your point of view, Mr. Blank, but why don't we look at it another way?" It is no good to win an argument and lose a sale. Thus, all sales interviews should be kept on a friendly, businesslike basis.

Early in an interview, some prospects will ask, "How much is this going to cost me?" A question of this nature reflects either:

(a) **A buyer's signal**, indicating that the prospect is sincerely interested in spending some money, or

(b) **An intended turndown**, his means of getting rid of you by replying, "Too much money; I haven't got the budget."

It is best to avoid mentioning price until you have told your entire story. To a prospect's pressing questions about cost, you can answer, "This idea is going to make money for you. In the long run, it will cost you nothing."

Salewise, it is smart to encourage the prospect to talk about himself and his business. From this, you can gather some very revealing clues to his thinking, and learn points of your sales talk to emphasize and which to soft peddle. Always demonstrate to a prospect that you respect his position. Even though many business people think of themselves as just "one of the boys," it is unwise to try to reach them on a buddy-buddy basis. Take it slow in using first names or nicknames. And dropping a "sir" into your presentation can only help.

Out in the business world there are some hard types that require special handling. **Old Stoneface** is one of these. He is the character who listens to your sales story without saying a word, nodding or shaking his head, and with no change of facial expression whatsoever. To break the ice with this tough customer, put questions to him, and lots of them, forcing his involvement. Ask, for example, "Isn't that a good idea?" after you have made some statement with which he can only agree. Get him into the habit of saying "yes."

A Tomorrow Man is the prospect who, after listening patiently to your complete sales pitch, asks you to come back later. Just what his reasons are for procrastinating are not easy to determine. Yet if he is not ready now, chances are 100 to 1 that he will cool off considerably by the time you return.

If you cannot close the Tomorrow Man on your first interview, despite all efforts, be sure to make a **definite appointment** for the next meeting. On your return visit, never ask him if he has made up his mind. An opening question like this gives him a chance to say that yes he has, and he has decided against buying an advertising schedule. It is best to open your remarks by indicating that you expect a favorable decision. "Just let me touch on a couple of important points we talked about last time," you begin cheerfully. Then you warm up the Tomorrow Man by repeating your entire sales presentation.

Mr. Know-it-All is the wise-guy prospect who imagines himself an absolute authority on advertising. He has the bad habit of interrupting your pitch with a variety of objections and observations, and he treats you like a kid who isn't dry behind the ears.

Prospects on such an ego trip can be hooked by their own vanity. Just agree with this prospective sponsor up to a certain point, and then say, "Your ideas are certainly good ones. Now please **tell me** just how we can best make a success of this venture?" The know-it-all sometimes falls solidly into the trap, and commits for a schedule more costly even than what you had in mind.

"Sorry, I can't see you today," are the words of **Mr. Busy-Busy**, a man of many excuses. "I'm going into a meeting"... "This is the worst possible day to talk to you"... "I'm much too busy." Tell this active gentleman that your merchandising program is very important, and will take about 15 minutes of his time, and that you wish to make an appointment in which to outline it. It is far better to come back than to try to force an unwilling prospect to listen to you after he has said he is too busy to do so.

In selling it is very important to reach the man who can say yes—the boss. But sometimes, it happens that you have to make your pitch to a subordinate, a **Nowhere Man.** The N.M. will listen to your story, because he has been asked to do so by his superior, but he can make no final decision and will have to take your proposal to the head man eventually. In doing this he is likely to forget important elements of your pitch. After you leave from this sales call, a follow-up is important. Send a letter to the subordinate, in which you review your proposition in writing, and mail a carbon of this letter to the boss. By doing this, you keep your proposal alive in both of their minds.

In conclusion, it must be pointed out that many salesmen fail because they simply do not **ask for the order.** As strange as it may seem to you this is a common failing. I know of a radio salesman who called on a jewelry store owner and presented a terrifically good plan. When the salesman finished outlining his program, the jeweler remained silent. The salesman then started his pitch all over again, and when he wound up his sales talk for the second time, the jeweler, who appeared interested, still said nothing. Finally, the salesman asked, "You seem to believe in this campaign, why haven't you bought it?" The prospect answered, "Because you haven't asked me to."

SIGN ON A SPONSOR'S DESK

In this office flattery will get you somewhere.

LOOK BEFORE LEAPING

We will take for granted that you have now found your first job in broadcasting. We will also take for granted that along the way in your career you will come across some dangerous problem areas that should be bypassed if you are to keep straightened up and flying right.

One of the first things to steer clear of is "announceritis." This affliction, so common among beginning performers (and some veterans) may otherwise be described as a **swelled head**. Symptoms of announceritis are a strutting walk, a cocked eyebrow and an undercurrent of "See how great I am!" running through everything you say and do, both on the air and off.

You need this kind of attitude like a large wart on the end of your nose. No points are scored with either your colleagues in the industry or station audiences by assumed superiority. To gloat over past successes is just as fatal as brooding over defeats. The football adage that "last week's headline won't win this week's game" surely applies.

Early in the game make up your mind that you are going to be a Regular Joe, putting out your best efforts, and accepting bouquets or brickbats as they come along, without either taking off on an ego trip or blowing your stack. Controlled confidence is your desired objective.

There are some radio announcers who have an astonishing faculty for disengaging themselves from their commercial copy. They get their kicks by actually making it a point to have their faces laugh while their voices continue to speak seriously. They horse around just for the hell of it. For professional reasons, it will do you well to shun these wise-guy characteristics. Such facial-vocal acrobatics take away—however minutely—from your delivery. As mentioned previously, selling comes first!

In broadcasting, as in other walks of life, are what we call **floaters**. These are the announcers and salesmen, particularly, who seem to be continually on the move, leaving a station in one locality to follow their trades in other places, merely for the sake of variety. A reputation as a floater is something you can do without, for station managers are especially reluctant to hire anyone of the "here today, gone tomorrow" nomadic tribe. Be sure you get started on the right foot by staying put in your first job for at least a year while you soak up knowledge, build a solid foundation for your career, and establish for yourself a good reputation as a dependable person.

Floaters seldom achieve anything worthwhile; they are too busy moving about. And since travel is very expensive, these gypsy workers never seem to have an extra buck in their pockets and are constantly in need of ready cash.

While we are on the important subject of money, it is necessary to bring up a couple of monetary bug-a-boos connected with your employment at a broadcasting station. Station managers dislike hearing, most of all, continuing requests for salary advances. Some workers, for reasons that have always been a puzzlement to me, are never able to stretch their paychecks from one payday to the next. As regularly as sign on and sign off, they appear once each week in front of the bookkeeper or manager begging for a few bucks to tide them over. Nothing bugs management more.

Going into your first job, and throughout your career, decide that you will avoid this aggravating practice, even if it necessitates hocking your watch or selling a gallon of your blood. And, for heaven's sake, try to be satisfied with the financial conditions of your employment. Management definitely does not care to hear repeated appeals for salary raises, and for talent fees for commercial work when such remuneration is not part of the station policy. In time you will earn the money—if you roll up your sleeves and do a good job.

One of the most common mistakes a disc jockey makes is carting home record albums belonging to the station, or "borrowing" tools from the engineering department, none of which are ever returned. In a word, this is stealing, even though the jock may not think of it as theft. Just don't do it; that's the honest policy.

Most all broadcasting stations, unfortunately, are battlegrounds for in-house feuds and petty jealousies. You will want to stand apart from these.

"Jock, the Wave Maker" is a familiar figure in the industry. He is a poisonous individual you should give a wide berth to. Jock seems to enjoy a strange satisfaction by making trouble. His principal interest in conversation is in damning the station, the management and the advertisers. His entire mental attitude is in thinking up sharp and sarcastic things to say.

Sometimes Jock is a very brilliant, sophisticated fellow whose cynical remarks about everybody and everything are so clever that it is fun to hear him sound off. But, bear in mind, that a person is judged by the company he keeps. In short, stay away from the wave maker, or you will most likely wind up unpopular and unemployed.

Another quick route to the unemployment line is by running up bills around town—buying a car, clothing, or a stereo set, or borrowing from a loan company—and then finding yourself "over your head" in debt and unable to pay them off. Station management is fast in firing a broadcaster who creates an unfavorable image of the company.

Management further frowns on employees who approach advertisers directly asking for discounts, credit, or other special deals. The sales department may have spent months, or years, developing a fine relationship with a local merchant, only to have this goodwill sabotaged by a finagling dodo.

The local starting point for a broadcaster who wishes some kind of product discount is to go to the sales manager direct and tell him what's on your mind. He knows the practices of the local business people best of all, and can probably get you a far better deal than you could arrange for yourself.

Someone once defined the **ideal guest** as "**someone who stays at home.**" This witticism is particularly applicable to visitors in broadcasting stations—a buddy of yours, a girlfriend, member of your family, or fans who find your work "fascinating." Without being aware of it, these guests can throw your entire presentation on the air off key by engaging you in conversation.

Other members of the staff do not care for strangers in the station getting in the way. After all, a broadcasting station is a place of business and not a recreation room. You should never permit **anyone** to loiter in the control room or studio. Concentration on your job comes first!

A lot of personal telephone calls are not going to enhance your popularity either. Keep these to the absolute minimum. Follow a strict, self-imposed rule, and give your **full attention** to your job during the hours when you are on duty.

The broadcaster most likely to succeed is the individual with the well-groomed look of the professional who is not only neat in appearance, but in production habits around the station. A slob-on-the-job seldom makes it past the boondocks. A first-class station demands a first-class person, not someone who comes on duty unshaven and wearing a soiled sweatshirt, faded jeans and dirty sneakers.

Laxity in grooming usually carries over into general studio etiquette. In general, it is the unkempt broadcaster who is the sloppiest in keeping logs, filing records and tapes, and leaving half-filled coffee cups and overflowing ashtrays behind.

At all times you must dress and act correctly if you intend to attract ad agency producers, advertisers and others who can further your professional status. Although there are shortcuts to getting started as a disc jockey or television performer, there are no shortcuts to professionalism.

Starting out your career by making serious mistakes that can be avoided doesn't make sense. It's like throwing away a Playboy calendar just because it's the end of the year.

SEX AND THE SINGLE DISC JOCKEY

The 1972 suspenseful movie, "Play Misty For Me," starred actor Clint Eastwood as a virile disc jockey plying his trade at KRML, Carmel-By-The-Sea. In the film Clint received a telephone request every night for the song "Misty" from a psychotic female fan, played by Jessica Walter. One thing led to another. Clint eventually met Jessica and made love to her. She became possessive. Then when he tried to discard her, Clint found himself with an insane murderess on his hands.

Although the film was entirely fictionalized, it illustrated dramatically for young disc jockeys a practical lesson in career conduct. Broadcasting is a glamorous profession that attracts women and girls. Air personalities become exciting celebrities to a certain segment of American womanhood, and opportunities for romantic indiscretions with these glamour-caught souls are wide open for a male performer looking for action.

Adolescence is a time for worshipping living idols, and teenage girls frequently get crushes on a radio performer. To them he is the closest thing to a celebrated movie star. These star-struck girls come to a broadcasting station in droves to demonstrate their admiration and reverence. The cookies and cakes that they bring to the DJ are oftentimes not the only gifts they are willing and eager to part with.

Many lonely married women, tired of bland diets of PTA meetings and bridge games, and of their husbands, find themselves spoiling for a little love. To prove to their egos that they are still attractive to men in their later years, they seek out a popular jock with whom to have an affair. Like a Dutch

Uncle, I must advise you that, in all instances, it is wise to keep hands off your admiring fans. Otherwise, you may be headed for a heap of trouble.

Once upon a time there lived in the Los Angeles area a married disc jockey whose show was broadcast each night from 9 until midnight. The jock had a very jealous wife who suspected him of playing around, but couldn't prove it. She kept close surveillance on his comings and goings.

The wife clocked his departure from their home at 6, and telephoned him at the station several times before his air shift to put her mind at ease that he was on the premises preparing for his program. She then listened to her husband's show for three hours, and at the end of it, clocked the minutes it took him to return home. To her way of thinking, his chances for unfaithfulness were nil.

But, in fact, the jock **did** have a girlfriend, and he visited his paramour several nights each week. This seeming miracle was accomplished electronically by the DJ, who thought himself exceptionally clever. With the assistance of a station engineer, **the jock taped his shows in advance!** Many of the programs that his spouse listened to were pre-recorded.

All went smoothly for the DJ's extra-marital love life until that fateful evening when the engineer accidentally erased the playboy's tape, and, by necessity, called in an off-duty announcer to carry on. You could say conservatively that it all "hit the fan" for the indiscreet jock.

Then there was the handsome New York kid with the "bedroom voice" who became the Don Juan of jocks in a little Southern town. On the air, during daytime hours, the turntable lothario romanced the women of the farming community with romantic dedications and suggestive patter:

"Time for our morning exercises, girls," he crooned. "Lie right down on your backs. Spread your palms about four inches to the side of those lovely thighs and we are going to firm up so nicely for your hubbies and boyfriends. Okay, loves, now keep your precious legs straight and shut tight. Real tight together. Just like the doors of Fort Knox. Now when I start counting, I want you pretties to bring up your right thigh just an inch or two. Just enough so I could slide my hand in there. You'll soon feel the difference it makes."

The ladies at home were spellbound by the DJ, caught in his web of words. It was a kind of hypnosis for them.

Had the talented young jock confined his techniques to the microphone, he might have gone on to great fame. Temptation, however, presented itself on every side. Within a few weeks, he was a familiar visitor to a variety of boudoirs and at the motels on the outskirts of town in the company of ladies, many of whom wore wedding bands.

Tar and feathering is an old-fashioned punishment that is seldom imposed anymore, but, in the case of the "smooth talker from New York," it was revived by a group of outraged husbands. One moonlit night the surprised disc jockey was captured by a party of hooded figures. He was stripped to the skin, plastered with hot tar and showered with feathers. The jock was deposited roughly on the edge of town and admonished. "Don't ever come back, you hear!" He never did.

A third example that I think well worth mentioning concerns the ill-advised activity of a former night jock of a Colorado station. One evening the program director paid an unexpected visit to the station. Upon entering the place he noticed that the control room was unoccupied and that a long-playing album was spinning on a turntable.

Where was the announcer? In the men's room, maybe.

The PD clicked on the lights in his office to a startling scene of the on-duty jock and a visiting female on the top of his desk thrumming together like rubber bands.

The badly-shaken announcer was fired forthwith, and he vanished into the night with the embarrassed girl, while the PD took over the disc jockeying chores for the remainder of the shift.

To say the least, an amorous DJ can find deep trouble by playing around with the girls of the town. An interesting sign posted in the control room of an Eastern rock and roll facility surely says it all:

WHEN NOT ON THE AIR PLEASE KEEP THE FOLLOWING CLOSED:

Your microphone

Your big mouth

Your fly.

PRANKS FOR THE MEMORY

A dignified Oregon newscaster appeared on the screen. He said, "Good evening. Here now is the news." A pet parakeet that someone had brought into the studio suddenly flew into the picture and settled comfortably on top of the newsman's bald head. The studio crew could be heard chuckling in the background while the newscaster, viewing himself and his fine feathered friend in the monitor, continued reading, but his delivery was punctuated with errors. The director of the show speedily rolled into the first commercial, and the bird was taken away.

What happened? Well, the bird either acted on its own initiative or, more likely, it was coaxed into the performance by the TV station's star practical joker.

In a Pennsylvania radio station two off-duty announcers zeroed in on a colleague who was reading a long list of want-ads-of-the-air while seated at a table in the middle of the main studio. Realizing that the announcer couldn't interrupt his broadcast, they calmly removed his shirt, slacks, loafers and socks and left him there in his peppermint-striped shorts.

"Be sure to see what's going on in Studio A," they told a merry group of sight-seeing college coeds in the waiting room. "It's a very unusual performance."

My first experience with practical jokers in broadcasting occurred during World War II when I was stationed at Armed Forces Radio Station WXLC in Dutch Harbor, Alaska. I was reading a 15-minute commentary, and was about two minutes into the copy when the door to the tiny announcing booth opened and a grinning engineer appeared.

As I kept on reading, the engineer caught my eye with a small vial of colorless liquid, which he held threateningly above his head, pretending it was nitroglycerin. All of a sudden, he smashed it in a metal wastebasket, startling me half witless. Then he exited quickly.

What he left behind was a stench bomb, and one of the foulest-smelling ever concocted by fun-loving chemists. By the end of the commentary, I was sure that my nose must have receded entirely into my face. It took days for the small room to freshen up sufficiently for human habitation.

All in all, broadcasting is a fun business. Thousands of serious-minded broadcasters have been subjected to pranks and shenanigans since the first station went on the air. On the TV screen, chairs have suddenly collapsed beneath stuffy performers. Copy from which radio announcers were reading has been set on fire by jokesters. Phony continuity has been substituted for the original just before airtime, and halfway through reading it an announcer has found himself mouthing senseless sentences.

Comedian George Lemont remembers when a prankster prop man put quick-drying cement in place of the vanilla pudding he was supposed to consume on camera. George, of course, couldn't spoon it, let alone eat it, as called for by the script.

Helpless in front of the cameras, Lee Grioux once struggled to open a lid of a cedar chest that had been nailed shut.

Disc jockey Jim O'Neil giving his all on the remote broadcast of a demolition derby in Eugene, Oregon, was asked by a security guard, who had been coached by staff members of Radio Station KEED, to interrupt his narration long enough to make an announcement. A 1970 automobile was parked illegally outside the arena, blocking all traffic. O'Neil read aloud the make, model and license number with all the proper flourishes, then realized it was his own car!

The most elaborate practical joke I know of was perpetrated on radio-TV personality Russ Coglin by the studio staff at KJEO-TV in Fresno. Unknown to Russ, all of the clocks in the station had been set forward 15 minutes. Thus, when his good friend Del Gore, the program host, began the live interview they were not actually on the air, although Russ thought so.

After introducing Russ to the imaginary audience, Del asked, "You meet a lot of lovely young movie actresses in Hollywood, don't you?"

Cautiously, Coglin answered, "A few."

"Do you sleep with any of them?" Del queried.

Coglin blanched. In a flash, his gaze swept to the camera. The red light was on! In the monitor, he saw a close-up of his own shocked face. Visibly shaken, he managed a very weak, "No, not me."

"Come on, Russ," Del pursued relentlessly. "Tell us the truth. You're not the kind of fellow to pass up an opportunity to make out with a beautiful, well-stacked broad. You do sleep with them, right?"

Coglin thought that Gore had lost his mind completely, and he might have fled the studio had not the host and the entire technical crew suddenly doubled up in uncontrolled laughter at his discomfiture. When the show actually went on the tube a few minutes later, Russ Coglin was still perspiring freely, and his performance lacked much of its usual spontaneity.

Not all on-the-air mishaps that bring guffaws behind-the-scenes can be blamed on practical jokes. Some of them come about unexpectedly. Steven Allen was once called upon by an advertising agency to take a hammer on live television and strike a fibre-glass chair as hard as he wished in order to demonstrate the non-breakable virtues of the new product. On the first whack, Allen smashed a hole right through the chair, bringing a roar of laughter.

A daily variety show on a Midwest station one day featured a "Noah's Ark" event in which the viewers were invited to bring in their unusual pets. Into the studio were brought monkeys, snakes, turtles and most every kind of animal, bird and fish imaginable.

All went well until about mid-show when somebody's pet hawk, Jacob, caught sight of somebody's pet rat, Charlie. Jacob flew boldly over, caught up the frightened rodent, winged to a high perch in the studio, and proceeded to eat Charlie in full view of the audience.

One of the longest sustained backstage yocks in the history of broadcasting happened on a late-night talk show to a noted actor who had, one year, been voted one of America's best-dressed men. The distinguished guest emerged from the wings (directly from the men's room) with his fly down, right to the last zipper notch. There is absolutely no way for a man to maintain his dignity once he discovers that his fly is open in front of millions of viewers and that it must remain that way until the first commercial break.

For sure, broadcasting provides some mammoth belly laughs for its people. A half-pound of yeast may be slipped surreptitiously into the cake mix of a TV chef by a practical joker. A player may sneeze during a televised badminton game with his false teeth sailing across the net like a birdie. Ours is a funny, always exciting, down-to-earth business to be a part of.

HONESTY IS THE BEST POLICY
—CONCERNING YOURSELF

In this book I hope I have answered most of your questions about breaking into broadcasting. I have tried to answer everything except your most important one: **Should I give it a try?**

Only **you** can answer that. You will be getting into a sometimes wonderful, sometimes frustrating, always hysterical business. You will meet good folks, and some not so good. In fact, now and then you will come across some honest-to-goodness, genuine, gold-plated SOBs. But to balance it off, you will likewise make the acquaintance of some of the nicest people on this earth.

I know that the ideas offered in this book are sound and practical, because I have seen them produce results time and time again. But I know, too, how very difficult it is to follow any program of self-improvement that lasts over a considerable period of time, no matter how fully convinced you may be of its value.

You must decide truthfully if you want to be a broadcaster. Can you honestly qualify? What are your real chances? If you are entirely honest about your assets and find that you don't have the **stick-to-itiveness** to eventually meet the specifications required by radio and television, don't feel disheartened. Congratulate yourself, instead, that you are intelligent enough to realize that you haven't got the calling. And go onto something else. But, on the other hand, if you are **absolutely sure** that you possess that certain undefinable spark and the rigid resolve to make a success of it, don't let anything or anyone stop you from achieving your goals.

The doors to broadcasting are open to everyone who sets his jaw in determination and his personal sights on the stars. Minority groups are coming into broadcasting. Nobody seems to have a reliable statistic on the job-getting gains of black people, Mexican-Americans, Puerto Ricans, Orientals and others, but the frequency with which they are turning up on-camera is evidence to even a casual viewer that broadcasting has long ceased being a white man's club.

WWLP, Channel 22, in Springfield, Massachusetts hired Paul Caputo, a man sightless since birth, as a newscaster. Mute and physically-handicapped persons now conduct their own shows across the country successfully.

What's more, broadcasting opportunities are everywhere, even in prisons. The California Institution for Men at Chino in Southern California has established a broadcasting school and a radio station, KCIM, for inmates. The ultimate objective of this workshop is to turn out qualified announcers, newsmen, salesmen and engineers.

The basic step in the endeavor to determine whether you have what is needed to succeed in a radio-TV career is to **rate yourself** as accurately as you can. To do this, I suggest that you seriously check off in your mind the following attributes and other qualifications which will be needed:

DO I HAVE A STRONG AND EARNEST DESIRE FOR SUCCESS?

HOW CONSCIENTIOUSLY DO I FOLLOW MY PROGRAM OF READING ALOUD? MY VOICE EXERCISES?

DO I REGULARLY TRY TO OVERCOME SLOPPY SPEECH HABITS, AND LOOK IN THE DICTIONARY CONSTANTLY FOR WORD PRONUNCIATIONS?

HOW CLOSELY DO I LISTEN TO SUCCESSFUL ANNOUNCERS AND ATTEMPT TO EMULATE THEIR PROFESSIONAL STYLES?

AM I MAKING HEADWAY IN BLEEPING 4-LETTER WORDS, "YOU KNOWS" AND SLANG OUT OF MY EVERYDAY CONVERSATION?

BEFORE I SPEAK, DO I ALWAYS CONSIDER THE CONSEQUENCES OF WHAT I AM ABOUT TO SAY?

CAN I TAKE SUGGESTIONS AND INSTRUCTIONS WITHOUT BECOMING ANGRY, EVEN IF THEY ARE NOT TOO TACTFULLY GIVEN?

TO WHAT EXTENT DO I TAKE ADVANTAGE OF EVERY OPPORTUNITY TO SPEAK IN PUBLIC, OR TO JOIN DRAMATIC ORGANIZATIONS?

IF A SERIOUS DISAPPOINTMENT COMES, CAN I FACE THE FACTS TO FIND A SOUND AND EMOTIONALLY COOL WAY TO REMEDY THE SITUATION?

WILL I SERIOUSLY FOLLOW MY SELF-HELP PROGRAM FOR A BROADCASTING CAREER COME HELL OR HIGH WATER?

If your answers to the previous questions convince you without a doubt that you are definitely turned on to get into the broadcasting act, here's hoping that you will become one of the greatest stars when **you're on the air!**

Index

A

Abolin, R. W. "Ozzie"	3
Accounting department	129, 134
Administrative assistant	129
Advertising agencies	59
Allen, Steve	22, 197
AM	60
Amateur Hour	33
Ameche, Don	87
American Federation of Television and Radio Artists (AFTRA)	140, 189
Amos and Andy	33
Amsterdam, Morey	154
Anchorman	119
Announcer	75, 107
Announcer-director	100
Application	100
Armstrong, Jack-the-all-American boy	35
Arquette, Cliff	149
Art director	111
Artists	144
Audio engineer	91, 117

B

Baby Snooks	33
Bailey, Pearl	21
Baker, Kenny	130
Barker, Bob	197
Baruch, Andre	33, 35
Bell, Lou	116, 119
Bell, Janis	129
Blue, Ira	27
Bookkeeper	82
Booth, Shirley	195
Bridges, Lloyd	85
Broadcasting	36
Broadcasting magazine	136
Broadcasting yearbook	36
Broadcast schools	37
Burnett, Carol	13, 193
Burns and Allen	33
Burr, Raymond	13
Business manager	129
Buzzi, Ruth	157

C

Cain, Bill	117
California Story, The	173
Call letters	56
Cameraman	93, 117
Campbell, Glen	21, 153
Campbell, Tom	52
Candid Camera	142
Cantor, Eddie	33
Captain Zig Zag	172
Caputo, Paul	218
Carlin, George	157
Carney, Art	133
Carney, Don	27
Carol Burnett Show, The	18
Carpenters, The	144
Carson, Johnny	13, 142, 167
Carter, Boake	177
Casanova, Frank	106
Casting director	133
CATV	134
—source book	136
Cavett, Dick	13, 170
Chacon, Rigo	125
Chaffetz, John	121
Chief engineer	67
Child, Julia	16, 193
Churchill, Winston	32
Clark, Jack	186
Clooney, Rosemary	87
Coglin, Russ	185, 197, 216
Collins, Dorothy	130
"Combo" men	65
Conrad, William (Cannon)	195
Continuity	130
—writer	74

220

Conway, Tim	197
Cooney, Pat	138
Cooper, Jackie	152
Copywriter	131
Cosell, Howard	154
Crane, Bob	52
Crane, Les	26
Crist, Judith	191
Croft, Mary Jane	152
Cronkite, Walter	25
Cullen, Bill	25

D

Dating Game, The	196
Daytimers	83
Dean, Dizzy	15
Dear Abby	173
DeLuise, Dom	30
Detter, Reynolds	117
Diller, Phyllis	13, 142, 197
Director	91, 103, 134
Director of photography	113
Disc jockeys	75
Douglas, Mike	13, 164
Downs, Hugh	53
Draper, Rusty	196
Duff, Howard	157
Duncan, Sandy	29, 137

E

Eastwood, Clint	212
Ed Sullivan Show, The	173, 197
Edwards, Ralph	195, 197
Ellison, Douglas	193
Engineering	
—department	116
—license	62
Ewbanks, Bob	13
Ewing Sam (Bud) Jr.	135

F

Fabray, Nanette	56
Fan club	200
Fedderson, Don	197
Federal Communications Commission	57
—offices	64
Feldon, Barbara	137
Fenneman, George	186
Film department	115
Floaters	209
Floor director	93, 106
FM	60
Fonda, Henry	87
Foster, Bud	106
Francis, Anne	130
Francis, Arlene	195
Frederick, Pauline	74, 191
Frost, David	56, 164, 199

G

Gabor, Eva	13
Gabor, Zsa Zsa	142
Galloping Gourmet, the	13
Gambling, John	185
Garagiola, Joe	27, 155
Garroway, Dave	142
General manager	95
Get Smart	137
Gifford, Frank	154
Gilliland, Allen T.	91, 136
Girl friday	55
Gleason, Jackie	13
Godfrey, Arthur	13, 26
Gore, Del	109, 126, 216
Gowdy, Art	13, 123
Gowdy, Curt	154
Graham, Virginia	193
Graves, Peter	162
Greene, Lorne	195
Gregg, Jan	161
Griffin, Merv	13, 164
Griffith, Andy	13
Grioux, Lee	24, 216
"Grip"	133
Gunsmoke	196

H

Hal	139
Hale, Barbara	193
Halla	139
Hamer, Vicki	129
Hannas, Art	168
Hart, Clarence	117
Harvey, Paul	52, 195
Haulman, Bob	123
Hayward, Don	121
Hee Haw	165
Heidt, Horace	153
"Help wanted" ads	37
Hokus Pokus	172
Horn, Woody	129
Hosfeldt, Robert	95, 99
Howard, Bud	115

221

Hurley, Lee	36
Hurley, Lu	169
Hyer, Martha	13

I

IBEW	116

J

Janssen, David	87
Jourdan, Louis	87

K

KABC	170
KAPY	176, 177
KAYO	57
KAZA	126
KCIM	218
KCOP	139
KDKA	56
KEED	216
KEEN	123
Kennedy, George	195
KFI	56
KGO	26, 169
KHJ-TV	138
KICK	57
KING	57
KING-TV	119
KJEO	115
KJEO-TV	216
Klein, Bob	117
KLOK	96
KMOX	33
KNBC	57
KNBR	173
Knotts, Don	167
KNX	46
KOLD	57
KOMO	157
KOOL	45
KOOL-TV	153
KOPO	107
KORN	57
Koufax, Sandy	154
KPHO	127
KPIX	115
KQIN	55
KRON-FM	144
KRON-TV	149, 170, 195
KROW	142
KRML	212
KSFO	29
KSJO	98
KSRO	30
KTIP	121
KTKT	106
KTVU	106, 147
Kubek, Tony	154
KVAL	104
KVOO	195
KXRX	125

L

LaCosse, Fred	119
LaLane, Jack	14
Lang, Kelly	170
Lange, Jim	186, 199
Lamm, Warren	111
Lawrence Welk Show, The	126
Lemmon, Jack	195
Lemont, George	149, 215
Let's pretend	195
Lewis, Hal	185
Lewis, Shari	113
Liberace	197
Lighting engineer	117
Linker, David	139
Linkletter, Art	52, 195
Little theatre	156
Logs	57
Loran, Darlene	102
Lord, Jack	18
Lovern, Leroy	117
Lowry, Terry	144
Ludlow, Don	115
Lynly, Carol	159

M

MacMurray, Fred	153
MacRae, Gordon	142
Mack, Ted	153
Major Bowes	33
Martin, Bob	117
Martin, Dean	200
Master control	117
Master control technician	91
Mayberry R.F.D.	196
Mays, Willie	16
McCarthy, Dan	127
McGee, Frank	56
McGovern, Gene	126
McIntyre, John	195
McKee, Gerry	117
McKuen, Rod	52, 197
McMahon, Ed	18
McNair, Barbara	13

Mead, Margaret	193
Meade, Julia	13, 159
Meyerson, Bess	13, 159, 193
Mike Douglas Show	135
Miller, Loren	175
Miller, Ann	87
Miller, Marvin	14
Moellering, Jan	101
Monitor	173
Moore, Mary Tyler	56
Morgan, Henry	186
Morrise, Wayne	138
Moulten, Jim	176
Mowbray, Barbara	55
Mowbray, John	55
Murphy, Jack	30
Murray the K	186
Music librarian	144
Myrt and Marge, The story of	35

N

Networks	59
Newman, Tom	117
News -department,	119
-director,	73, 119
NEWS-TV	123
Newton, Wayne	153
Nordstrom, Gene	119

O

Ohran, Paul	165
O'Neil, Jim	216
Operations manager	98
Original Amateur Hour	173
Ott, Pam	128
Owens, Gary	199

P

Paar, Jack	52, 170
Page	127
Palmer, Betsy	186
Pamphilon, Clay	106
Park, Stew	98
Peale, Norman Vincent	197
Penner, Joe	33
Phillips, Wally	185
Photographer	132, 144, 163
Plato, Armand	156
Prentiss, Paula	157

President Roosevelt's Fireside Chats	33
Princess Pat	172
Producer	132
Production Coordinator	101
Program -department	59
-director	72
Promotion & public service	102
Putnam, George	185

R

Radio station formats	61
Randall, Tony	195
Rayburn, Gene	142
Reagan, Ronald, Governor	196
Reasoner, Harry	56, 196
Rebello, Bill	117
Reimers, Ed	186
Reporters	163
"Reps"	59
Rickles, Don	16
Risinger, Jim	104
Rivers, Joan	142
Ranger Roger	172
Roberts, Al	117
Robinson, Edward G.	87
Room 222	196
Rosemarie	197

S

Sales -department	59, 79, 126
-manager	59, 81
Salesman-announcer	83
Sample, Alvin "Junior"	165
Sanders, Marlene	191, 195
San Jose State	95
Sarge	195
Scully, Vince	123
Sesame Street	173
Shadow, The	35
Shore, Dinah	193
Simmons, Sylvia	123
Singin' Sam	15
Smith, Barbara	129
Sommes, Jack	137
Sorkin, Dan	199
Spann, Owen	199
Sportscaster	73
Sports director	121
Spot announcements	59
State, San Jose	99
Station manager	71

PN
1991.55
.E9

Steele, Ted	142
Stern, Bill	144
Stevens, Connie	156
Story, Ralph	199
Stoopnagle and Budd	33
Strange Lands and Seven Seas	139
Stringer	162
Studio announcer	108
Sullivan, Ed	15, 168
Swayze, John Cameron	17
Swift, Allen	87, 153
Sykes, Brenda	196

T

Talent agencies	140
Technical director (TD)	91, 117, 134
This Is Your Life	195
Thomas, Lowell	31
Today Show	27
Tonight Show	18
Traffic	82
-department	59, 128, 134
Trotta, Liz	191
Truth or Consequences	123, 173, 197
Tufts, Sonny	138

U

UHF	87
"Uncle Don"	27
Unger, Bill	126

V

Vallee, Rudy	87, 153
Vanderbilt, Amy	13
VanDyke, Jerry	167
VanVoorhis, Westbrook	15
Viacom Systems	135
Vera, John	127
VHF	87
Video engineer	134
VonTobel, Rudolph	53
VonZell, Harry	15

W

WABC	57
Waggoner, Lyle	18
Wallace, Mike	195
Wallington, Jimmy	19
Walter, Jessica	212
Walters, Barbara	144, 193
Warren, Earl	30
Wayne, David	87
WCBS	57
Weaver, Charley	166
Weaver, Dennis	13
Webb, Jack	168
Wells, Orson	13, 195
West, Adam	52
West, Barbara	196
WGN	56
WHAM	57
WHDL	168
White, Betty	56, 191
WHIZ	57
Widmark, Richard	195
Wilkens, Bob	147
Williams, Andy	21
Willman, Ken	119
WIND	45, 57
Winters, Jonathan	16
WJBK	130
WKCR	123
WLW	33
WLWC-TV	121
WMAQ	46
Women's director	73
Wonders of the World	139
Woodnall, Joanne	129
WOR	27, 56
WOW	167
WQBC	15, 175
Wrightson, Earl	142
WSB	56
WTTW-TV	121
WTVN-TV	130
WTVU-TV	193
WWL	33
WWLP	218
WXLC	215

Y

Yearwood, Jack	127
Yochim, Bob	126
Young Widder Brown	35
Your Hit Parade	

Z

Zimbalist, Efrem Jr.	

PN
1991.55
.E9